Global Energy Shifts

Global Energy Shifts

FOSTERING SUSTAINABILITY IN A TURBULENT AGE

Bruce Podobnik

 Temple University Press
PHILADELPHIA

Temple University Press
1601 North Broad Street
Philadelphia PA 19122
www.temple.edu/tempress

⊗ The paper used in this publication meets the requirements of the American National Standard for Information Sciences—Permanence of Paper for Printed Library Materials, ANSI Z39.48-1992

Library of Congress Cataloging-in-Publication Data

Podobnik, Bruce, 1968–
    Global energy shifts : fostering sustainability in a turbulent age / Bruce Podobnik.
        p.   cm.
    Includes bibliographical references and index.
    ISBN 1-59213-293-6 (cloth : alk. paper) — ISBN 1-59213-294-4 (pbk. : alk. paper)
    1. Energy policy.   2. Power resources.   3. Energy industries — History. 4. Sustainable development.   I. Title.

HD9502.A2P63 2005
333.79—dc22

                                                    2005043943

2 4 6 8 9 7 5 3 1

Contents

List of Illustrations

Acknowledgments

There are many people who assisted me in completing this study. First, I would like to sincerely thank Beverly J. Silver for her support. It was while working with Professor Silver at Johns Hopkins University on a research project examining patterns of labor unrest and capital mobility that I initially formulated my research topic. Not only was she of crucial assistance during the earliest stages of my research, but Professor Silver provided guidance throughout all subsequent stages as well. I extend my gratitude to her for assisting me at every step of my work.

I would also like to thank Christopher Chase-Dunn for helping me bring this project to completion. An opportunity I had to collaborate with Professor Chase-Dunn on a project examining dynamics of hegemonic rivalry allowed me to develop my understanding of world-systemic patterns of geopolitical competition. Professor Chase-Dunn also provided particularly insightful advice on the final version of this study.

There are many other people who helped me finish this book. I would like to thank Giovanni Arrighi, David Harvey, Frederick Buttel, Helmut Anheier, Melvyn Kohn, and two anonymous reviewers for commenting on different versions of this study. I am also grateful for the support of my editor, Peter Wissoker, who remained patient as I worked on revisions. For providing friendship and support, I would like to thank Mirella, Nhan, Mahua, Melanie, Rick, Kari, Tay, Amy, Larry, Mia, Yona, and everyone at the Zen Community of Oregon. And, finally, I am truly thankful for the support of my family. I wish my mom, Judy, could be here to read this book, but I am happy that my dad, Don, and my sisters, Kim and Kerri, will finally be able hold this book in their wonderful hands.

Global Energy Shifts

1 Global Energy Shifts in World Historical Perspective

In the latter part of the nineteenth century, the citizens of Great Britain faced what seemed to be a bleak energy future. Commentators argued that the country's most important energy resource—coal—was destined to run out within a generation or so. At the same time, they argued that there was no viable alternative to coal. Two primary solutions to Britain's perceived energy crunch were therefore offered: Military strategists were urged to undertake expeditions to seize control of coal reserves in foreign lands, and companies were urged to drive their workers harder to increase domestic production of the resource. But these efforts were met with resistance from other colonial powers and from unions inside Britain. Meanwhile, cities across the nation grew increasingly choked with the sulfurous pollution flowing out of factories, railroads, and homes.[1]

Accounts at the time argued that Britain was destined to lose its global preeminence as its coal reserves disappeared and as social and environmental problems originating from the industry tore at the fabric of British society. In the eyes of many commentators, the resource that had once fueled the rise of the nation's fortunes had begun to contribute to the weakening of the British Empire.

Shifting forward in time, intriguing parallels between the "coal panics" that swept through Britain in the latter part of the nineteenth century and the "oil panics" that grip the world today are clear. The world's oil infrastructure is threatened by insurgencies in many countries, and there are widespread fears that reserves of oil will be unable to meet world demand within a decade or so. At the same time, many analysts claim that there are few viable alternatives to oil. Likewise, the recommendations that flow from this view rather eerily echo those proposed in the nineteenth century. Governments across the world are again being urged to seize military or commercial control over key oil reserves, and local officials in many countries are being urged to remove constraints on domestic oil extraction to maximize production.[2]

But resistance to these efforts is even fiercer today than it was in the nineteenth century. Oil-producing countries like Saudi Arabia, Iran, and Venezuela are protecting their independence from external pressure. Meanwhile, social chaos is engulfing countries like Iraq, Nigeria, and Russia and undermining their ability to provide oil to the wider world-economy. Given these constraints, competition for oil reserves is escalating between large consumers like the United States, the European Union, and China. Underlying these pressures is a host of growing environmental dangers. From deteriorating local air quality to the many dangers posed by climate change, there is widespread consensus in the scientific community that overreliance on fossil fuels must soon end if ecological catastrophe is to be avoided.[3]

If long-term stability is to be achieved, significant reforms must clearly be made in our global energy system. But can fundamental changes really be made in time to avoid severe strategic, commercial, social, and ecological crises? Although this is a question that cannot be definitively answered, the historical record shows that remarkable changes in the global energy system have occurred in the past, which should allow for some cautious optimism that similarly profound transformations can be achieved in the future.

Consider, again, the example of Britain. With hindsight we can see that nineteenth-century analysts were wrong to claim that there were no alternatives to coal. How were they to know that a new resource—oil—would ease Britain's energy crunch? After all, it played almost no role in providing energy in that society in the 1870s and 1880s. But by the beginning of the twentieth century, the British government and private industry were spearheading a shift toward increased reliance on this new resource. This transition was given added urgency as social and environmental tensions emerged around coal sectors. As a result of these intersecting forces, by the end of the First World War oil had become the fastest growing source of energy in Britain.

This shift toward oil was repeated in country after country, beginning in the United States and spreading to countries as diverse as France, Russia, Japan, and Australia. In fact, within a few decades the first global energy shift of the modern era—toward increased reliance on oil—was well underway. And even as the world was rocked by the First World War, the Great Depression, and the Second World War, investments in oil-based technologies and infrastructures continued doggedly ahead. Massive new fleets of ships, trucks, cars, and airplanes were built, and entire regions of the world were incorporated into a new energy system based on oil. The speed and magnitude of this shift, carried out during very turbulent decades, was quite breathtaking.

Current claims that little can be done to shift away from overreliance on oil sound rather defeatist in light of the transformations that were achieved in the global energy system during this earlier period. Just as many nineteenth-century analysts were unable to conceive of changes that would bring oil flooding into a system dominated by coal, many contemporary analysts seem unable to imagine conditions that could foster the rapid growth of alternative energy systems and bring new forms of renewable power streaming into a system that is currently dominated by fossil fuels. They argue that alternative energy sectors are not yet viable or that the capital does not exist to fund major new infrastructures. These claims overlook the achievements of the early twentieth century, when new financial investments were undertaken in the toughest of times that propelled the development of new energy industries forward.

Although there can be no guarantees about what will happen in the coming decades, there are many reasons to believe that shifts of the magnitude and speed achieved in earlier eras can again sweep through the global energy system. As the historical analysis carried out in this study will show, entirely new systems of energy production, transportation, and consumption have repeatedly enveloped the world in a period of about fifty to sixty years. This remarkable history of transformation should give concerned citizens the confidence to move forward with efforts to reform the energy foundations of the contemporary world.

An Empirical Overview of Global Energy Shifts

In this study, an energy system is defined as originating from the naturally occurring, primary resource that is harnessed for human use. Primary energy resources include, for instance, wood, coal, oil, natural gas, radioactive energy, running water, wind, and solar power.[4] An individual energy system, then, is defined as the interconnected network of production, transportation, and consumption that delivers one of these specific energy resources to people for use in their daily lives. The global energy system, meanwhile, is the even more complicated totality of these individual energy networks. As will become clear in the coming chapters, events in one energy system can have reverberating effects throughout the other systems and can even cause long-lasting transformations in the structures of energy production and consumption on a global scale.

Over the long arc of human history, the relationship between primary energy resources and technological systems has become increasingly complex. Today it is often difficult to determine the source of the energy that powers the machines we use. Consider, for instance, the case of electricity. Although electricity occurs naturally as lightning and static, these natural forms cannot easily be captured for human use. Technologies have therefore been devised to transform a variety of primary energy resources—like coal, natural gas, nuclear power, and running water—into electricity. These different streams of electricity are then fed into a utility grid, which distributes energy to any light, appliance, or motor connected to the network. Because electricity is clean and quiet at the end-point of consumption, its origins in coal, oil, natural gas, and uranium are hidden from view.

One way of piercing through the complexity of modern energy industries and getting at key underlying features of the global energy system is by focusing an analytic lens on the primary resources that are being harnessed for human use. Looking at the primary energy resources that are being fed into the world's energy grids will render visible important features of the global energy system that may otherwise be obscured.

For this reason, the point of departure of this study is an investigation of patterns in the extraction and consumption of primary energy resources. From this perspective, an energy shift is defined as the process whereby a new primary energy resource is harnessed for large-scale human consumption. This incorporation may occur through the creation of new technologies, or through the resource being fed into preexisting systems. But whatever the intermediate process might be, the underlying material fact is that a new source of energy is being captured for use.

Adopting this resource-based perspective allows for a straightforward depiction of successive waves of incorporation of energy over the modern period. The first modern energy system, based on coal, grew steadily in the nineteenth century and reached maturity in the twentieth century. The second modern system, based on oil, expanded rapidly during the twentieth century and is now reaching maturity. Systems based on natural gas and hydroelectricity are also in the process of attaining global reach. Meanwhile, nuclear power industries have expanded in certain regions of the world, but they are not likely to be capable of becoming truly global in scale. Finally, systems based on renewable energy sources such as wind and solar power are beginning to gain regional importance, although they are not yet near attaining a global scale.

Figure 1.1 shows shifts in the percentage of the world's commercial energy provided by the most important primary resources over the last two centuries. The figure demonstrates that, in 1800, most of the world's commercial energy came from wood and other biomass materials. Of course, since these biomass resources tended to be harvested and consumed in small-scale operations, data on their use have to be treated with caution. Once we turn to coal and other industrial resources, though, the empirical record becomes more reliable.[5]

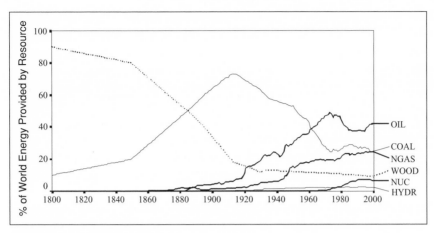

FIGURE 1.1. Global Energy Shifts, 1800–2000
Sources: See Appendix A.

According to data compiled by historians, coal went from providing approximately 10 percent of the world's commercial energy in 1800 to over 60 percent in 1913. This shift to reliance on coal was very rapid when compared to the time it took for preindustrial societies to harness the power of resources such as wood and wind. Still, the speed of the shift to coal pales in comparison with what came later.

It took over a hundred years for the coal system to mature to the point where it was providing 50 percent of the world's commercial energy supplies. The shift to reliance on oil was much more rapid. In 1910, oil provided only about 5 percent of the world's commercial energy supplies. Sixty years later, though, oil was supplying approximately 50 percent of the world's energy. Natural gas has also undergone a rapid process of growth; whereas it provided only about 6 percent of the world's commercial energy in 1946, by the year 2000 it was supplying about 24 percent (about the same as coal's contribution). A central goal of this study is to uncover the factors that allowed for the rapid expansions that have occurred in these fossil fuel industries.

Not all commercial energy industries have undergone such rapid and sustained trajectories of growth. Nuclear power experienced some expansion in the 1970s and 1980s, but by the late 1990s it had reached a plateau at about 6.5 percent of primary world energy supply. Energy from hydroelectric facilities, meanwhile, underwent slow growth throughout the twentieth century, so that by the year 2000 hydroelectricity was providing about 3 percent of the world's commercial energy. Meanwhile, all modern

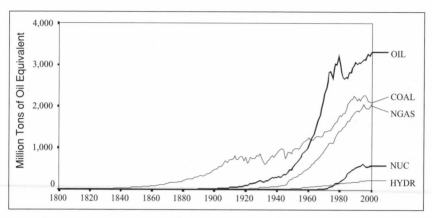

FIGURE 1.2. World Commercial Energy Production, 1800–2000
Sources: See Appendix A.

renewable energy systems (including wind, solar, geothermal, and modern biomass) combined provided only about one half of 1 percent of the world's commercial energy in the year 2000. This is, of course, a sobering statistic for anyone concerned with the environmental viability of modern society.

Another way to map these successive waves of incorporation is by looking at the volume of energy being absorbed into human societies. To create this kind of figure, though, we first need to transform the wide variety of energy resources we are examining (tons of coal, barrels of oil, and so on) into a common unit. This analysis follows many other energy studies by using one metric ton of oil equivalent as the common unit. Taking the amount of energy contained in one average metric ton of oil as a basic unit, it is a straightforward operation to calculate how many of these energetic units are contained in a given quantity of coal, natural gas, or any other resource.[6] Once this transformation is carried out, we can create a graph that depicts the volume-based growth of primary energy resources over the modern period on the basis of a common metric.

Figure 1.2 provides this alternative representation of global energy shifts. The volume of coal production (measured here in tons of oil equivalent) grew steadily up to 1913, entered into stagnation until 1945, and then resumed modest growth in subsequent decades. Oil production, on the other hand, is marked by exponential growth in the period 1945–1973 and then by volatility since that time. For its part, natural gas has achieved rapid and steady growth since the Second World War. Nuclear power output achieved significant growth in the 1970s and 1980s, but then plateaued. Hydroelectric power production has maintained a slow but steady growth rate since the early 1900s. Meanwhile, modern renewable energy systems provide such a tiny proportion of the world's energy that they do not even register in the figure.

There are a number of important observations to be made about the patterns of evolution in energy industries that emerge from these empirical pictures. What is perhaps most remarkable is the speed with which huge volumes of energy can be incorporated into the world-economy. Oil went from providing about 44 million tons per year in 1910 to around 3 billion tons per year in 1970. This phenomenal rate of increase in oil production has almost been matched by the expansion of natural gas. In fact, from 1940 to 2000, natural gas output rose from 76 million to over 2 billion tons of oil equivalent per year. Given the right constellation of circumstances, massive volumes of new resources can clearly be incorporated into the global energy system in a handful of decades. If these new resources can be integrated into preexisting infrastructures and commodity circuits, then the rate of absorption of new resources can be accelerated even more.

It is important to point out, though, that these new energy systems have been superimposed on top of older systems that continue to expand. The shift toward greater reliance on coal that occurred in the nineteenth century, for instance, was overlaid on a world that was using ever-growing quantities of wood. Similarly, the shift toward increased dependence on oil and natural gas that took place in the twentieth century was layered on top of a still-growing coal system.

The energy shifts that have occurred in the modern era have therefore been relative rather than absolute shifts. That is, while a higher percentage of the world's energy is now coming from oil and gas, coal consumption continues to grow in absolute, volume terms. Overall, global reliance on hydrocarbon resources has increased exponentially throughout the modern era. In fact, today coal, oil, and natural gas resources combined provide approximately 90 percent of the world's commercial energy requirements. At some point in the future, however, an absolute shift will have to be achieved where systems based on coal, oil, and natural gas are replaced by something else. Since coal is very plentiful, pressures will always exist to revert back toward greater reliance on this very polluting resource. With careful preparation, though, a shift toward cleaner, more environmentally benign energy systems can be achieved.

Achieving a shift away from hydrocarbons and toward modern renewable energy resources may seem impossible from today's perspective. However, it is clear that the global energy system has gone through important periods of profound, nonlinear change in the past (Smil 2003). The key challenge is to develop an understanding of what caused these earlier transformations, and then foster a similarly profound shift in the coming decades.

The conceptual tools of world historical analysis will illuminate the underlying societal dynamics that have generated profound transformations in the global energy system in the past and will continue to do so in the future. How this particular research tradition can help make sense out of the nonlinear transformations that are etched in the historical trajectory of the global energy system is described below.

A World Historical Interpretation of Global Energy Shifts

One of the most important research traditions in the social sciences centers on the effort to come to an understanding of the historical origins, contemporary dynamics, and likely future trends of the capitalist world-economy that has come to incorporate the entire globe. While it is readily acknowledged that there are always elements of unpredictability in human affairs, a broad number of scholars argue that there are discernible patterns of evolution that underlie the development of this world-economy. It is hoped that, by coming to a better understanding of key trajectories of societal evolution operating within the system, our ability to intervene in intentional ways to reform problematic aspects of our globalizing world can be enhanced.

The scholars who have most clearly articulated this approach operate within what has come to be known as the world-systems perspective. This perspective was first introduced in an explicit way by Immanuel Wallerstein in his influential book *The Modern World-System* (1974). Drawing on theorists such as Marx, Weber, and Braudel, Wallerstein argued that a capitalist world-system came into existence in Europe in the sixteenth century. Following its creation, this world-system expanded to incorporate ever-wider geographical areas in intensifying networks of political, economic, social, and cultural interaction. Once incorporated into the system, nations, industries, or communities could no longer be treated as independent units. They instead had to be seen to be heavily affected by changes in the broader world-system, and their actions likewise often reverberated throughout the system.

Over the last twenty-five years, researchers from such disciplines as sociology, political science, economics, and history have contributed to the effort to uncover patterns of continuity and change in the evolution of this world-system.[7] From all the insights generated in this world historical research tradition, four are particularly relevant to an analysis of global energy shifts. The first is that world events in general, and global energy transitions in particular, are driven in part by a dynamic of geopolitical rivalry that fluctuates between periods of intense and moderate conflict. The second is that there is a process of corporate competition that also alternates between periods of radical industrial innovation and periods of more predictable growth. The third is that dynamics of social conflict likewise go through alternating phases of radical and moderate intensity.

The fourth is that these systemic dynamics of geopolitical rivalry, commercial competition, and social conflict interact in a process known as the hegemonic sequence. The modern world has repeatedly alternated between periods of relative order, during which a great power (or hegemonic state) is able to impose stability across the globe, and periods of chaos, in which powerful countries compete for dominance. During periods of relative order, international conflict is contained, commercial prospects are enhanced, and social unrest is more or less suppressed. Periods of chaos, however, are characterized by warfare, economic crisis, and radical outbreaks of social conflict.

Transitions from periods of world order to world chaos profoundly impact the global energy system. As this study will demonstrate, periods of relatively linear, predictable growth in global energy systems are achieved during times when a hegemonic state is able to contain dynamics of geopolitical, commercial, and social competition within moderate frameworks. Conversely, periods of more profound change in global energy systems occur when hegemonic stability breaks down and the pressures of warfare, economic crisis, and social conflict can no longer be contained.

This, then, is a brief summary of the key conceptual tools used in this study of global energy shifts. Below is a more detailed discussion of the ways in which each of these systemic dynamics individually impact energy industries, as well as the way they periodically interact to produce global energy shifts.

Geopolitical Rivalries and Global Energy Shifts

One of the most enduring features of the modern world-system has been that of geopolitical rivalry among nation states for military and economic primacy.[8] Since long-term military strength is intimately tied to the economic health of a given state, political leaders have often intervened in commercial matters, especially in sectors that are judged to be of strategic significance. This state intervention has been particularly common in the case of energy industries.

The efforts of empires and nation states to stimulate domestic energy production and to gain access to foreign energy resources have a very long history. As early as the fifth century B.C., the city-state of Athens employed force to compel its hinterlands to provide wood resources for energy and construction purposes. This state-enforced exploitation of wood grew during the Roman Empire, while authorities throughout China, India, North Africa, and Western Europe also came to intervene in circuits of wood production and consumption in the classical and medieval periods.[9]

In the modern period, state intervention in energy sectors has in some ways waxed and waned. During the Napoleonic Wars, for instance, state agents in Britain and France engaged in many efforts to increase their own coal output while simultaneously attacking enemy mining operations. State intervention in coal eased once these wars came to an end, only to intensify again later in the nineteenth century as coal-powered industries became critical to modern war machines. This interventionist dynamic was further strengthened during the First and Second World Wars, when key adversaries struggled to maximize their access to coal and oil supplies for military campaigns. The current era is witnessing a renewed intensification in state intervention as powerful nations across the world struggle to secure access to affordable oil resources.

Although there has been a shift between periods of strong and moderate intervention, states have rarely retreated all the way back to laissez-faire policies in times of peace. Instead, there has been an expansion of state mechanisms for influencing energy industries. Early state action was limited to, for instance, seizing private coal reserves for use in battle or conscripting laborers to work in mines. By the eighteenth century, however, governments began using a variety of regulatory, fiscal, and procurement tools to facilitate the expansion of domestic energy operations. In the nineteenth century many governments also started funding research and development programs designed to improve the operation of specific technologies and to further the scientific understanding of energy combustion processes. By the twentieth century, leading states were regularly utilizing a range of overt and covert strategies to secure access to key energy supplies. And in recent decades, multilateral agencies such as the World Bank and the International Energy Agency have developed some capacity to influence energy industries and markets on the international level.

We can trace these dynamics of state intervention in energy sectors thanks to important work carried out by a broad number of historians and social scientists. McNeill's (1982) analysis of the industrialization of warfare, for instance, is crucial for comprehending the growing degree of state intervention in modern energy industries. Meanwhile, Fine (1990) and Fremdling (1996) provide good descriptions of government influence in coal sectors, and a host of scholars have examined state interventions in the international oil system. Of particular note are the studies offered by Yergin (1991) and Nowell (1994).[10] More recent work by Bunker and his colleagues[11] and by Harvey (2003) and Klare (2004) demonstrates that states continue to intervene extensively in energy industries in the current period.

As important as state support is for emerging energy systems, the historical record shows that governmental interventions alone are not enough to foster the full maturation of an energy technology. In fact, this study will demonstrate that dangers emerge when state officials try to push energy industries forward irrespective of commercial and social resistance. The case of nuclear power reveals the danger of this kind of state overreach in particularly clear terms. Still, interventions by an era's most powerful nations have always played a critical role in successful global energy shifts. A future shift to more sustainable energy technologies will similarly require important levels of support from political authorities in many countries across the world.

Corporate Competition and Global Energy Shifts

Innovations in the world-economy are also propelled forward by competition among private firms for dominance in key commercial sectors.[12] During the modern era few arenas have been as profitable as those that provide energy resources to industry and consumers. As a result, the global energy system has long attracted the attention of entrepreneurs and private corporations. The resulting dynamics of commercial competition have had a profound impact on the global energy shifts that have taken place in the modern period.

Over the last two centuries, commercial energy industries have tended to shift from situations of intense corporate competition to ones of controlled, oligopolistic competition. Most coal mining regions, for instance, first saw the emergence of many small companies that engaged in cutthroat forms of competition. Over time, many of these regions went through processes of industrial consolidation that left a select group of companies dominant in regional or national industries. This shift from cutthroat to oligopolistic forms of competition was even more marked in the case of oil. A select group of oil companies were able to expand their operations to national and international levels, so that the entire oil system quickly became dominated by a small number of firms that worked together to limit intense competition. Both the coal and oil industries, however, witnessed periodic breakdowns in these industrial truces. The reintensification of corporate competition that ensued then helped set the stage for major shifts in each of these industries.

The rise of the multinational oil corporation stands as a remarkable case of entrepreneurial growth and innovation in the face of significant market challenges. As discussed in later chapters, oil prices were higher than coal prices in most markets when oil firms began their rapid expansion. And as oil companies began penetrating into established markets, coal companies tried to mobilize public opinion and government influence against oil. The oil companies that succeeded in winning markets away from coal clearly did so in extremely competitive environments.

The ability of oil companies to grow in the midst of challenging market conditions highlights the capacity for commercial innovations to be achieved in what are often assumed to be static energy industries. The business history of energy shows that firms can overcome significant price hurdles and open up new arenas for market expansion and profit making. Understanding the forces that have allowed for major shifts in the global energy system requires examination of corporate competition and innovation in global energy industries.

Social Conflict and Global Energy Shifts

While political and commercial dynamics have long been recognized as major forces of global change, processes of social conflict have less often been accepted as being of similar significance. A growing number of scholars working within the world historical tradition, however, have come to argue that dynamics of social conflict have a central role in the evolution of the capitalist world-economy and the energy industries embedded within it.

It was, of course, Karl Marx and Friedrich Engels (1848) who most ardently maintained that human history is driven by processes of class conflict. More recent scholars, such as Hobsbawm (1962), Gordon (1980), and Mandel (1980), have lent added weight to these early claims. Recent work by Beverly Silver (2003), meanwhile, draws together historical evidence of both a qualitative and quantitative nature to demonstrate that the long-term impact of social conflict on the world-system has been as profound as the forces of geopolitics and commercial competition.

Dynamics of social conflict have had far-reaching impact on the historical evolution of large-scale energy industries. Struggles carried out by coal miners to protect their livelihoods and communities, for instance, changed work procedures, wage rates, and safety regulations in coal sectors throughout the world. Oil industries, for their part, have been transformed by labor and nationalist struggles with deep roots in civil society. Campaigns undertaken by environmental and indigenous rights movements have constrained the expansion of hydroelectric and nuclear projects.[13]

Over the course of the modern era, we can document a rise and fall in the social tensions surrounding energy industries in particular locations.

For instance, waves of labor militancy swept through coal industries in Europe and North America in the late nineteenth century and then again after the Second World War. Within oil, social tranquility was more or less maintained until the latter part of the twentieth century. But then workers and citizens in key oil-exporting nations began participating in campaigns to overturn corrupt regimes and nationalize petroleum industries. While phases of intense social conflict eventually gave way to times of compromise and moderation, they left lasting changes in their wake. Union organizations consolidated their influence in many coal regions, for example, while nationalist movements retained important degrees of control over domestic oil industries.

Over time the types of social movements impacting energy sectors and the strategies they employed have multiplied. Union campaigns have grown in sophistication, often coming to employ experts from the worlds of law, politics, and media. Nationalist movements have made even more successful use of political campaigns, while contemporary environmental and indigenous rights groups have used tactics like demonstrations, media campaigns, consumer boycotts, and legal challenges in their attempts to reform energy industries. In the current period, insurgent groups such as al Qaeda are using even more extreme tactics to destabilize operations and destroy energy infrastructures in critical regions.

Given the pervasive influence of social movements in the evolution of modern energy systems, it is surprising that the mainstream energy literature has so often treated workers and activists as irrelevant or passive agents.[14] This inattention to social dynamics of unrest is one reason why mainstream analysts have been frequently unable to forecast eras of radical change in global energy industries. One of the main goals of this study, then, is to advance our understanding of this previously underexamined dimension of global energy shifts. By documenting the various ways that social conflict has provoked periods of fundamental, nonlinear change in the past, we will be better able to understand how such dynamics are likely to operate in the future.

The Hegemonic Sequence and Global Energy Shifts

The systemic dynamics of geopolitical rivalry, commercial competition, and social conflict each partly follow their own trajectories. Indeed, as subsequent chapters will demonstrate, individual dynamics can at times move to the forefront in causing change in energy industries. However, it is my contention that for energy shifts to become truly global in scale, each of these systemic dynamics must be operating together in strongly transformative ways. This convergence of influences is most likely to occur at times of world crisis, when great powers are in decline and all facets of the world-system are in flux. Though the rest of this study fleshes out the argument in more detail, what follows here is an overview of this relationship between the hegemonic sequence, the interaction of systemic dynamics, and the consolidation of energy shifts on a global scale.

The British and American periods of hegemonic order each rested upon the growth of each era's key energy resources. Early successes in the pioneering of new energy systems gave each rising great power important advantages over its competitors. These resource advantages were important in determining which state would emerge victorious in times of war. Once a new world order had emerged, the careful distribution of energy resources fostered alliances and enhanced the perceived legitimacy of the political order maintained by the leading state.[15] During these periods of hegemony, dynamics of commercial competition and social conflict were also kept within moderate frameworks, which provided the stability required for the maturation of the era's key energy industries.

A central characteristic of human history, however, is that no world order has ever been sustained in perpetuity. Instead, hegemonic powers have consistently lost their technological and financial leads as new competitors have emerged. Dynamics of corporate competition have become more chaotic in periods of late hegemony, and declining hegemonic powers have tended to overextend themselves militarily in efforts to retain dominance. This has intensified social unrest at home and throughout the world. Ultimately, the ability of the declining hegemon to contain chaos evaporates, and a new period of intense, often violent interstate conflict ensues.

It is during these periods of world chaos that fundamental changes most readily occur in the global energy system. This study will show, for instance, that the first era of transition in energy systems in the modern period—the shift from coal to oil that took place during the mid-twentieth century—corresponds to a period of hegemonic crisis when all three systemic dynamics became radically transformative in nature. Geopolitical tensions escalated, prompting large-scale interventions by leading states in the accelerated development of oil systems. Private corporations, which had already begun investing in new oil industries, were then able to greatly expand the scale of their operations, and an intensification of labor conflict in coal industries throughout the world undermined profits and confidence in maturing coal sectors. Given these multiple pressures, the transition toward oil was remarkably rapid and far-reaching.

A second period of significant change in the global energy system, the crisis of the 1970s, similarly corresponds to a time of hegemonic crisis.

Weakened by its military campaigns in Southeast Asia, the United States temporarily lost its ability to contain threats to the international oil system. Corporate rivalries in oil markets intensified, and social unrest in key oil-producing nations exploded. These conditions then set the stage for a wave of nationalizations and price adjustments to sweep through international oil markets. However, in the early 1980s the United States and its allies reestablished their capacity to impose order in key oil sectors, which fostered a shift back toward primary reliance on the resource. Systemic turbulence declined in the face of this retrenchment, and the global energy shift toward more sustainable technologies stalled.

The collapse of socialist regimes in Eastern Europe, the rapid defeat of Iraq in the first Persian Gulf War, and strong economic growth in the United States and East Asia were taken by many as proof that the United States had begun a second phase of hegemonic leadership in the 1990s. However, most world-systems researchers argue that the United States has remained trapped on a trajectory of decline.[16] Their arguments appear particularly prescient today, following the series of terrorist attacks and international conflicts that have again called into question the stability of a world order led by the United States. Security threats have multiplied, financial crises have rippled through the world-economy, and military campaigns spearheaded by the United States have undermined the perceived legitimacy of this nation's global project in the current period.

This contemporary period of world disorder may escalate into catastrophic crisis. If that were to occur, the stability of the global energy system would, of course, be one of the least of our concerns. However, there may be a silver lining in these times of trouble. As the historical record shows, global energy shifts occur most readily during eras of turbulence and chaos. Indeed, as the conclusion of this study will argue, the current era of crisis provides a unique opportunity for governments, corporations, and communities across the world to mobilize in favor of a shift to a more sustainable global energy system.

Human Agency and Global Energy Shifts

My interpretation of global energy shifts emphasizes the central role of human institutions and conflicts in influencing the evolution of large-scale energy industries.[17] This is an argument that departs from those that draw on more deterministic models, which are not particularly helpful in shedding light on modern energy shifts so long as they ignore the ways in which societal dynamics foster large-scale changes in the global energy system.

Early research on the historical evolution of energy industries is marked by a strong current of resource determinism. In the formulations of Jevons (1865), Ostwald (1909), and Carver (1924), for instance, shifts from one source of energy to another were seen to have been caused mainly by the effects of resource depletion. While resource depletion has played a role in some energy shifts, however, modern transitions have not generally been associated with resource scarcity effects. For instance, most shifts toward greater reliance on oil have taken place in countries that still have abundant supplies of coal. A simple model based on resource depletion cannot account for these increasingly common shifts.

Another version of resource determinism focuses on the characteristics of specific resources and maintains that these qualities are what drive energy shifts. For instance, there is a widespread belief that physical features of oil—the fact that it is a liquid or that it burns well in small engines—are what determined its eventual triumph over coal. This interpretation, however, overlooks other characteristics of oil—the fact that it explodes or that it needs to be refined—that led many people to question its usefulness in earlier eras. Instead of winning converts because of its physical properties, oil-based systems had to be aggressively marketed before they attained popular acceptance. While the physical characteristics of energy resources certainly play some role in their diffusion, analyses that focus exclusively on this dimension are prone to neglect the societal dynamics that more profoundly influence large-scale energy shifts in the modern age.

In raising questions about narrowly deterministic perspectives, I am not attempting to entirely discount the impact that resource endowments have on transformations in the global energy system. Each has a role to play in setting the material constraints within which human institutions operate. As new energy technologies are developed, material constraints are at least temporarily eased. But if available resources fail to keep up with growing demand, these constraints tighten. Over the period 1800–2000, material constraints were never rigid enough to propel large-scale energy shifts on their own. Instead, competitive dynamics within human societies were the central forces driving transformations during an era of abundant energy. But energetic constraints on human societies have not always been loose, and they are not likely to remain so in the future.

In premodern times, technological innovations occurred relatively infrequently and they added only modestly to the energy supplies available for human use. As a result, consumption generally pressed close up against the limits of resource availability. Indeed, there were cases in which premodern societies overtaxed their resource base, leading to intensified social conflict and even civilizational collapse.[18]

During the modern era, the exploitation of fossil fuel reserves allowed energy production to greatly outstrip established consumption requirements, and material constraints on the global energy system therefore became quite loose. We are now, however, entering a new phase in which fossil fuel consumption will need to be scaled back—first for environmental reasons, and then because of depletion effects—while consumption continues to grow rapidly. As a result, the global energy system is again entering a period of tightening resource constraints.

One intriguing thing about the current period is that resource constraints are drawing tighter just as the capacity for human beings to alter energy trajectories is growing stronger. In the past, social groups have generally intervened in uncoordinated ways. The history of energy shifts, therefore, has often been a tale of the unintended consequences of human intervention. Today, though, coordinated efforts are being carried out on many levels to devise solutions to our energy crises. Political elites are beginning to construct international agreements and institutions designed to facilitate the global expansion of renewable energy systems. Multinational energy corporations are increasing their investments in renewables to spread their risk and position themselves for success in emergent sectors. Environmental groups are intensifying their efforts to contain the most damaging kinds of energy projects while pushing for a more rapid expansion of renewable energy technologies. These coordinated efforts provide hope that a future global energy shift can be achieved even more quickly than those that have occurred in the past.

When analysts in nineteenth-century Britain surveyed their energy options, the horizon of possibilities they could consider were much more limited than ours are. Today we can find inspiration in the fact that a new energy system, based on oil, spread across the world during the most chaotic and violent decades of the twentieth century. This provides hope that new energy systems, based on renewable technologies, can begin similar trajectories of growth even in the turbulent era that lies ahead.

There are, of course, no certainties when it comes to forecasting what will happen in the global energy system. But it is important that concerned citizens across the world forge ahead with the knowledge that fundamental changes can be achieved. Just as the gloom of late-nineteenth-century Britain gave way to unexpected new energy horizons based on oil, the current era of crisis can similarly give way to a new era of hope as renewable energy industries are encouraged to replicate rapid expansion cycles. Advocates of this kind of future global energy shift need not be utopians. We can instead proceed with a historically grounded, realistic, and yet ambitious agenda of transformation in the coming decades.

2 The Rise of Coal

A t the beginning of the nineteenth century, communities across the world relied on resources like wood, wind, and water for most of their energy needs. Only about 10 percent of the commercial energy used in 1800 came from coal. Within a matter of decades, however, a massive growth in coal mining had taken place. Coal fueled profound transformations in industries across Europe and North America, but it also helped propel new waves of colonial conquest and global trade. In fact, during the nineteenth century coal allowed the capitalist world-economy to undergo a process of expansion that fundamentally refashioned societies all around the globe.

But what factors allowed this first global energy shift of the modern age to take place? And why did this coal system become centered in Western Europe, rather than in the Asian societies that had more experience using the resource? As shown in this chapter, unique dynamics of political and commercial competition emerged in Europe that allowed technological barriers to the large-scale use of coal to be overcome. Transformations in financial institutions and social relationships generated by an emerging capitalist system also allowed for the development of large-scale coal industries in Europe and North America. The order imposed by British hegemony then fostered the expansion of coal mining across the world.

Under the intersecting influence of these societal dynamics, coal grew into the world's most important energy system in the course of only 60 years. An analysis of this historical transformation reveals the capacity for the world to undergo far-reaching global energy shifts in a handful of decades.

The Early Development of Coal-Based Technologies

It has often been argued that the main factor pushing societies toward coal was the material constraint posed by deforestation. It is true that wood scarcities in Europe prompted many efforts to develop new energy sources in the early modern era. However, an analysis of the factors underlying the large-scale shift toward coal that began in the eighteenth century reveals that resource depletion pressures did not by themselves produce the transition. Instead, specific dynamics of geopolitical and commercial competition found in Europe were of central importance in propelling a shift to the large-scale use of coal resources.

It is important to begin by noting that the human use of coal has deeper historical roots in Asia than in Europe.[1] In fact, the earliest evidence of the human consumption of the resource comes from Bronze era archeological sites in China. By the eleventh century coal was being used extensively in Chinese metallurgical industries. The Chinese also appear to have been the first to dig shafts into the ground to extract coal, though in most cases the pits only went down to the point where water was encountered.

While the Chinese led the way in developing uses for coal, there is also a long history of coal consumption in Europe. By the classical age of antiquity it was being used to heat homes in Greece, for instance, while Roman soldiers used it to heat their outposts in northern Britain. In the eleventh and twelfth centuries, surface deposits were being mined in England, Belgium, and France. The development of underground shafts appears to have begun sometime thereafter, and by the thirteenth century efforts were being made in England to penetrate below water tables along particularly thick mineral seams.

These European efforts to engage in large-scale coal mining were provoked to some degree by problems of deforestation. The extensive use of wood for ship construction, iron ore production, and home heating led to severe deforestation around urban and industrial centers, especially in England. From the sixteenth through the early eighteenth century, the price of wood and charcoal rose across Western Europe. Efforts were made to harness energy from peat and from water, but in many areas the lack of resources became a barrier to industrial production.[2]

Although energy deficits were intensifying in many parts of Europe during the Renaissance, it was by no means obvious that coal could fill the gap. There were significant hurdles standing in the way of a widespread shift toward reliance on coal. The water that seeped into shafts posed a key problem, both to Chinese miners in the eleventh century and to European miners in the seventeenth century. Even when coal was extracted, its price was inflated by the difficulties faced in transporting the bulky material to consumers.

The challenges of pumping water and transporting coal were joined by important social and economic barriers to greater reliance on this resource.

It was difficult to find laborers who were willing to work in coal mines, for instance. Even if workers could be found, it proved hard to mobilize the financial capital needed to expand mining and transport operations. Finally, coal was widely regarded as being inferior to traditional wood products. It was usually harder to light than wood or peat, and it tended to generate more noxious smoke than wood. Most consumers therefore resisted using the material.

Therefore, although resource pressures were prompting a shift away from wood and toward something new, there were significant barriers in Europe that prevented a large-scale shift toward reliance on coal. A particular convergence of social and geographical factors was needed to allow for this transformation. That this convergence of factors occurred in Western Europe in the late seventeenth and early eighteenth centuries was partly a product of good fortune and partly a reflection of the social transformations being wrought by an emergent capitalist world-system.

As Marx, Weber, Polanyi, Wallerstein, and many other researchers have demonstrated,[3] it was in Western Europe that capitalist forms of production and social relations first succeeded in overturning feudalism. On the ground this was a slow and halting process, but it was given formal recognition by the signing of the Peace of Westphalia in 1648. Enshrined within the Treaty of Westphalia were two central principles that would play a crucial role in allowing the capitalist system in general, and the coal system in particular, to grow in an increasingly unfettered fashion.

The first principle stated that the separate political entities that existed in Western Europe at the time of the treaty each had sovereign rights over their respective territories. No longer was it accepted that an overarching authority, based on religion or dynasty, ruled the affairs of all European states. Instead, interactions between sovereign states were to be governed by the rule of law, as determined by negotiations between the states themselves. The second principle extended this concept of interstate law by stating that conflicts between sovereign states were not to be visited upon civilian populations in general, or upon commercial enterprises in particular. Under the terms of the treaty, war between sovereign states was to be permitted, but only insofar as it involved professional combatants on designated fields of battle. Exchanges between private citizens, entrepreneurs, and financiers were to be allowed and protected, even if their governments were at war.

The Treaty of Westphalia helped solidify the existence of two of the systemic dynamics I take to be of crucial importance in understanding global energy shifts.[4] The legal codification of the European interstate system helped ensure that the dynamics of geopolitical rivalry would persist over the following centuries. And the legal protections offered to financial interests helped ensure that dynamics of commercial competition would likewise thrive and expand. Of course, the treaty by no means called these dynamics into existence; it merely reflected the capitalist revolution that was already under way. And the treaty's protections were never precisely adhered to; there were repeated attempts to overturn the principles embedded in the treaty. However, these efforts failed, and from 1648 the dynamics of geopolitical and commercial competition became enduring features of an emergent capitalist world-system.

These dynamics had a profound role to play in the early consolidation of a coal-based energy regime. The first clear manifestation of their impact came with the development of the steam engine. This technological marvel would not only provide a solution to drainage and transport problems in mines, but it would also become the motive technology lying at the heart of the global coal system. In examining its development, we see that technological innovations do not arise and spread across society in a natural or automatic fashion. Their diffusion is instead fostered by the convergence of societal factors that occur at particular places and times.

Detailed sketches of hot air chambers from Asia demonstrate that inventors in China, India, and elsewhere had figured out how to harness the power of heated gases as early as the fourteenth century (White 1960). However, social factors did not foster a shift toward coal-powered steam engines in Asia; but in Europe, a number of conditions came together to spur the diffusion of the steam engine within coal industries.

Experiments with steam engines began much later in Europe than they did in Asia. By the thirteenth and fourteenth centuries, though, craftsmen in Italian and Germanic city-states were tinkering with various kinds of hot air chamber designs. These efforts seem to have been hindered by limited financial support from public or private sources, however, as well as by the difficulties of exchanging information in a context where wars often prevented international travel. As a result of their relative isolation, European engineers were repeatedly forced to reinvent the hot air chamber.

The post-1648 period, however, allowed scientists to exchange information across Europe. Moreover, state support for research into strategically useful technologies grew in many countries as contending powers strove to develop military advantages in a context of almost continual warfare. It was in this fertile context that the steam engine was developed.[5]

A key figure in this process was the English scientist Robert Boyle. Boyle began his career by traveling to Geneva and Florence, where he learned about earlier work on hot air chambers. Once back in England, Boyle and other leading scientists were commissioned by the British monarchy to found the Royal Society of England in 1660. Supported by funding from the central government, Boyle was able to work on high-pressure vessels while his colleagues carried out important work on a variety of other scientific problems.

A similar trajectory was followed by Christiaan Huygens. Huygens first established a reputation as a leading scientist through his work on hot air pumps in the Netherlands. He was then recruited by King Louis XIV of France to open the Academy of Sciences in Paris in 1666. It was in this Academy that Denis Papin, a Frenchman, sketched out plans for a steam engine in 1679. The first steam engine to be patented, by the English military engineer Thomas Savery in 1698, was based upon Papin's design.

Although the steam engine evolved out of a broadly international research effort, once it had been patented in Britain it became official policy to prevent the diffusion of the technology to other countries. For the first three decades of the nineteenth century, the British government used border controls and legal prohibitions to prevent the export of steam engines and technical drawings.[6] It was only in the 1830s, once Britain had solidly established its hegemonic position in the world-system, that restrictions on the diffusion of steam engines were relaxed.

Even with the imposition of export controls, strong dynamics of geopolitical and commercial competition ensured that steam engine designs spread widely across Europe. Some of these transfers were accomplished through legal sales, once export restrictions had been removed. But transfers also occurred through less legal channels. Entrepreneurs in England often dismantled engines and shipped them to engineers on the Continent, who then created unlicensed replications. State authorities in France, Prussia, Sweden, and Russia also financed espionage efforts to gain access to steam engine technology.[7] Meanwhile, autonomous efforts to improve upon basic designs took place in many countries. By the early nineteenth century, the technical capacity to build coal-fired steam engines had undergone wide diffusion across most of Western Europe.

It was not a coincidence that this technology emerged in the particular time and place that it did. Resource pressures played a role, though they were by no means the crucial element in spurring the development of the steam engine. Of more significance was the emergence of a set of conditions that fostered the public financing of research and interactions between scientists in different parts of Europe. Processes of geopolitical and commercial rivalry, linked to an emergent capitalist world-economy in Europe, were ultimately what created the conditions needed for the invention and diffusion of the steam engine. This innovation, in turn, set the stage for the consolidation of a coal system that would transform the world in the coming decades.

The Emergence of a Coal-Based System in Britain

A widespread expansion in coal production demanded more than just a single technological innovation. It required important social transformations as well. The shift to coal was contingent on the emergence of a new group of entrepreneurs and a new set of financial institutions that could mobilize large amounts of capital. It also required the creation of a new class of laborers who could be sent to work in the deep shafts that the steam pump now made possible.

The shift from traditional forms of coal mining to newer, steam-assisted mining occurred most rapidly in Britain. But why did the shift to modern forms of coal mining occur most rapidly in this country? To answer this question, we not only have to turn to an analysis of geographic factors—which are commonly offered as the key reason Britain jumped ahead in coal mining—but we also must examine crucial social transformations that were occurring in the country. Moreover, we have to locate the United Kingdom in its position at the center of an expanding world-economy during the eighteenth century.

Britain's advantageous participation in the Atlantic triangular trade, and its later extraction of wealth from India, had direct and indirect roles to play in stimulating rapid growth in British coal production. World-systemic processes were therefore closely linked to domestic transformations, and each helped foster the early shift toward coal that occurred in the United Kingdom in the seventeenth century.

Let us begin by looking at the traditional explanation for Britain's early shift toward greater reliance on coal, which emphasizes the roles of resource depletion and geography. Here we find that Britain faced a unique combination of pressures and advantages. Problems of deforestation were more intense in the British Isles than in the rest of Europe, and so the incentive to develop a new source of energy was stronger.[8] Moreover, many coal deposits were found near navigable waterways. As a result, in the eighteenth century it was possible to deliver large quantities of coal to urban consumers in the United Kingdom without having transportation costs drive the price of the resource up too much.

If the United Kingdom was favored in these material terms, it also benefited from specific transformations that were occurring in the social fabric of the country. The consolidation of capitalist practices in elite circles of British society provided a favorable financial context for mine expansions, while the expulsion of peasants from rural land generated the labor force required to work the shafts. Each of these social processes advanced more rapidly in Britain than in its Continental counterparts.

A class of entrepreneurs that had roots in Britain's traditional agrarian society, but also links to London financial houses, began to take shape in the late seventeenth century.[9] Their ability to undertake commercial ventures was assisted by the founding of the Bank of England in 1694, as well as the creation of regional banks and insurance companies in the following decades. By the mid-eighteenth century, these British businessmen were supported by a financial infrastructure that was far more advanced than what was available on the Continent.

This emergent class of British entrepreneurs was at first seen as a threat to the old aristocratic order, but by the mid-eighteenth century their social status had been solidified, thanks in large part to the growing contribution of trade and industry to the British economy. Their position in British society was also enhanced by the fact that these aristocratic entrepreneurs often rescued estates from ruin by opening new lines of business in rural areas.

These aristocratic entrepreneurs played a particularly crucial role in the modernization of coal mining in eighteenth-century Britain.[10] Many of the largest coal owners of the era, in fact, were estate owners who turned to coal mining in an effort to supplement declining agricultural incomes. Given their relatively privileged position in English society, these coal owners were often able to self-finance their operations. Others were able to turn to relatives or peers for personal loans. Still other coal owners, mainly in the northeast of England, were able to take out loans from regional banks or the Bank of England.

It is also important to note that Britain's position at the center of a growing colonial network provided direct and indirect support for its domestic coal industry. The most direct link came as merchants in coastal cities invested the profits they made on foreign trade into coal mines. Many of the Scottish merchants involved in the North American tobacco trade, for instance, invested profits into their local coal mines. They also supplied private capital for construction projects that expanded canals, which facilitated large-scale shipments of coal. Much the same thing occurred around Liverpool, where the merchant elite often reinvested their dividends from foreign trade in regional coal mines, canals, and port facilities.[11]

There was also an important, indirect benefit that accrued to the coal sector from Britain's central position in the world-system. The influx of profits from the Atlantic Trade, and the conquest of India, allowed the United Kingdom to pay down its international debt, which reduced domestic interest rates. This made it more affordable to raise capital for mining projects in England, as compared to its Continental counterparts.

A wave of new investments, which originated as profits from colonial trade, therefore began to flow into British coal mines and transportation infrastructures during the mid-eighteenth century. This new capital permitted a whole range of new technologies to be introduced into mine shafts across Britain. Meanwhile, roads, canals, and new port facilities were constructed to facilitate the movement of the coal to major cities.

In addition to the important changes that were happening at the pinnacle of the social order, the United Kingdom underwent a process of proletarianization in the mid-eighteenth century that fostered growth in the coal industry. For centuries, enclosures of rural land had been forcing peasants off communal lands and into new kinds of occupations. But in about 1750 the enclosure movement underwent a renewed phase of expansion. As a result, mine owners were able to gain access to a new pool of workers at a crucial point in the evolution of the industry.[12] Without this pool of labor, it is doubtful that Britain could have achieved the remarkable growth in coal output that was registered in the eighteenth century.

The serfs, peasants, and farmers who were forced into the mines encountered a uniquely harsh work environment. Whole families were put to work in very difficult and dangerous occupations. Young boys were generally set to work operating ventilation doors in the shafts when they were around ten years old. After about ten years of work underground, they were permitted to take their place at the coal face. There they joined more experienced male hewers, who cut passages through the rock in patterns designed to maximize coal removal while keeping roofs intact. Meanwhile, women and girls were put to work hauling coal and performing other support tasks in the shafts. Women also carried out the vital tasks of cooking, cleaning, and caring for infants in the communities that grew up around the mines.[13]

Armed with picks, dynamite, and steam pumps, miners were ordered to dig their way further into the ground than had ever before been attempted. With each new meter, the dangers mounted. Poisonous and flammable gases seeped out of the coal seams, often killing workers through suffocation or massive explosions. The wood beams used to reinforce mine shafts increased the danger of fire, while providing imperfect protection against cave-ins. Sudden encounters with underground rivers periodically flooded entire mining complexes. Clearly, mining in deep shafts required a workforce with great physical endurance and expertise if high levels of production were to be achieved in anything like a safe manner.

Given the escalating level of danger encountered in the mines, it is somewhat curious that there are few records of organized worker rebellion in coal regions in the seventeenth century. In part this can be explained by the compulsory forms of labor that were employed. In Scotland, for instance, serfs were often transferred from their lords' estates into the mines, while in northeast England many of the miners were bonded laborers. The mining communities were also quite isolated, and camp houses were generally owned by the local landowner. Workers and their families could be dismissed from the community for any infraction. The operators of pits often circulated lists of unruly workers to their counterparts across the region, so being dismissed from one pit often meant that a miner and his family had to move out of their native community altogether.[14] Given these constraints, the emergence of organized labor militancy had to wait for transformations in the mining industry and the political culture of Europe that would occur in the coming decades.

Intensifying Systemic Dynamics and the Expansion of Coal

At the end of the eighteenth century, Britain led the world in the production and consumption of coal. Over 70 percent of the horsepower generated by steam was located in the United Kingdom, while mining output in the country far outstripped that of its competitors. However, an escalation in the dynamics of geopolitical rivalry and commercial competition at the turn of the century led to the diffusion of modern coal mining practices throughout Europe. As a result of the interaction of these systemic processes, by the mid-nineteenth century most of Western Europe had undergone a significant shift toward reliance on coal for the satisfaction of its energy requirements.

From the perspective of geopolitical dynamics, the late eighteenth and early nineteenth centuries were dominated by the struggle between Britain and France for dominance over the world-economy. One of the early consequences of this struggle was an intensification of pressures on the French state. Rising expenditures on military campaigns forced the monarchy to increase tax burdens on an emergent bourgeois and urban working class, thereby contributing significantly to the outbreak of revolution (Skocpol 1979). As we shall see, the ideological effects of the French Revolution would ripple throughout the world in a matter of decades. Meanwhile, the Anglo-French conflict redoubled in the wake of the collapse of the revolution, as Napoleon marshaled an even stronger challenge to British power.

The resulting conflicts had important consequences for an embryonic coal system on the Continent.

During the Anglo-French Wars of 1793–1815, the military demand for iron in Britain and France grew steadily. Expanding domestic coal production or seizing enemy coal deposits therefore became an important strategic objective. The Saar coal fields in particular emerged as an important military asset. Located along the French-German border, the Saar was the sight of heavy combat. Napoleon captured the region in 1792, and his engineers made strenuous attempts to increase coal output. With the subsequent defeat of Napoleon, the Saar was taken over by Prussia and renewed efforts were made by state authorities to expand coal output in the region.[15]

Heightened state intervention in coal industries in many European countries dates from soon after the Napoleonic wars. Officials on the Continent, concerned with spurring industrialization for both military and commercial reasons, increased public investments in their mining sectors. This early state support was crucial, especially in a context where most private consumers had not yet begun purchasing coal in large quantities. By the first decades of the nineteenth century, Continental arms manufacturers and military commanders were signing large purchase contracts with coal companies, thereby stabilizing demand for the fuel. As a result of this state support, coal mines and their associated industries began to grow steadily in France, Belgium, and the Germanic states in the early nineteenth century.

French authorities, for instance, increased subsidies to domestic coal mines after 1815. They also financed the construction of the first coke-blast furnace in their country and helped fund the construction of a national railway system from the 1820s onward. The French navy also led the way in adopting armored steamships, which then stimulated a broader race to develop coal-powered naval fleets. Britain immediately responded with a large increase in public funding for steam ships. Subsequently, the Crimean and the Franco-Prussian wars demonstrated the effectiveness of utilizing railways and steam ships in combat. By the mid-nineteenth century, virtually all major European states were heavily subsidizing their coal mining and transportation systems.[16]

Following successful demonstrations of steam-based land and naval systems, state subsidies for railroads and ships increased throughout Europe and North America. The period from 1880 on marks the emergence of the first true military-industrial complexes, with European, North American, and even Japanese companies entering into long-term development projects with army and naval contractors.[17]

This state support was crucial to the expansion of coal sectors outside of Britain. Coal mines on the Continent generally faced greater hurdles to expansion than those in the British Isles. Many mines in France, the Germanic states, and Belgium were located far from major transportation routes. Meanwhile, financial institutions were less developed on the Continent than in Britain. Indeed, the Byzantine nature of Continental banking law at the time often forced coal entrepreneurs to spend more time seeking capital from acquaintances or regional nobility than engaging in actual mining. In such a context, family and personal ties to state authorities were key to gathering the capital required for expanding mining and transportation systems in the region.[18]

Even in the face of these constraints, however, coal sectors began to grow in Belgium, France, and the region that would become Germany. In each case, state-sponsored programs of industrial promotion were crucial to advancement. Over time each country also witnessed the consolidation of a class of mining magnates who were able to create modern business organizations and introduce new production techniques into their operations. As a result, the Belgian coal fields of the Borinage, the Prussian mines of Silesia and the Ruhr, and the French mines of Nord/Pas de Calais all witnessed steady growth from the 1820s onward.[19]

The material achievements of this mutually supportive combination of public and private investment in coal were quite impressive. As revealed in Figure 2.1, the United Kingdom continued to dominate the world in coal output during this period, but the rest of Europe also witnessed an acceleration in coal output after 1830. As a result, during the first sixty years of the nineteenth century, European coal production rose from around 18 million metric tons to 125 million tons, which represents a huge incorporation of new energy into the core of the world-economy.

This continent-wide energy shift involved the construction of a massive new industrial infrastructure, consisting of steam engine and boiler manufacturing plants, railroads, iron ships, and a host of other associated industries. To get an idea of the magnitude of this accomplishment, consider the following. The first iron railroad was opened in England in 1825 (linking the coal mines of West Durham to the port of Stockton). By 1860 over fourteen thousand kilometers of rail had been laid in Britain, eleven thousand in the Germanic states, nine thousand in France, and almost two thousand in Belgium (Cipolla 1976). Overall, in less than sixty years a massive new industrial infrastructure based on coal had been deployed throughout Western Europe and parts of Eastern Europe.

Clearly, early capitalist societies were capable of orchestrating a rapid expansion of a new energy system. Those who hope for an equally rapid

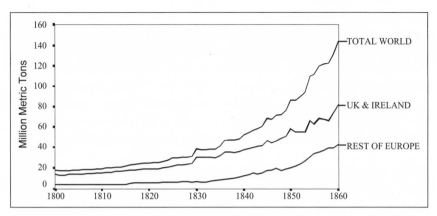

FIGURE 2.1. Expansion of the Coal System, 1800–1860
Sources: See Appendix A.

expansion of renewable energy systems in the current age can certainly take heart from this early demonstration of the speed with which new energy systems can attain continental diffusion, even in rather inhospitable circumstances.

Coal and British Hegemony

The global energy shift toward coal that occurred during the early nineteenth century not only transformed societies within Europe, but it also had far-reaching global consequences. This energy shift became intimately associated with a new process of conquest that forcibly incorporated new regions into an expanding world-system. Coal-powered ships and railroads allowed Britain and its Continental rivals to seize control over territories in Asia, Africa, and the Middle East that had long resisted conquest. Coal-driven transport systems then allowed for a radical increase in the volume of goods moved from the periphery into the core of the world-economy. The construction of these coal-based systems in the colonial world was also used to legitimate rule by European powers, since they were heralded as manifestations of progress and modernization. By the mid-nineteenth century, geopolitical, commercial, and social interests fostered the emergence of mutually self-reinforcing cycles between colonization and coal consumption that accelerated the globalization of this new energy system.

The country that most benefited from this new phase of coal-assisted colonization was Great Britain. Having defeated France in 1815, Britain established itself as the uncontested hegemon of the expanding world-system. Over the next six decades, expansions in British colonial holdings

were intimately related to its strong domestic coal industry and technological capacities. The use of coal-based technologies initially allowed
for significant improvements in manufacturing capacities within Britain.
Coal-based systems then facilitated the imposition of military control
over areas that had long resisted colonial rule. These systems also allowed for the development of lucrative, British-controlled bulk trading
networks. Finally, by selectively integrating colonial peoples into new,
coal-powered commercial networks, the British fostered the development
of classes of comprador elites that helped legitimate and stabilize colonial
rule. As this process of colonial expansion and commercial deepening occurred, the shift to coal-based systems underwent a process of truly global
diffusion.

One of the first demonstrations that there was a link between colonial
expansion and coal-related technology came in 1842, when a small number
of British gunboats succeeded in defeating a much larger fleet of traditional
Chinese vessels during the First Opium War. The ability of steam-powered
gunboats to maneuver quickly in both coastal and river waters proved to be
of decisive importance in these battles. British naval officers adjusted their
tactics quickly, shifting from primary reliance on sailing ships to steam
gunboats in their campaigns against the Chinese (Headrick 1981). Over
the next decade and a half, the British made effective use of gunboats to
destroy the Cantonese fleet and penetrate upriver to attack Beijing. They
also began to employ gunboats in campaigns in Burma, India, and the
Middle East. France, Belgium, and Germany quickly adopted similar
strategies. By the 1870s France was using steam boats in Indochina, while
Germany and Belgium relied on steam boats to penetrate up rivers into
the African interior. Coal-powered vessels therefore played a key role in
fundamentally shifting the balance of military power between Europe and
even the strongest Asian and African societies.

Once regions had been subjugated, the ability of British state authorities
to introduce steam-powered machines into colonial domains then became
an important ideological component of their rule. The coal-powered
railroads that were constructed in India, for instance, were touted by
the British as manifestations of the progress and advancement they were
bringing to their colonial subjects. These messages resonated with local
comprador elites who were allowed to benefit from the new circuits of
trade the railroads created. By selectively granting access to local trading
surpluses, British authorities were able to stabilize their control over the
Indian subcontinent. With the cooperation of these local elites and the further expansion of rails and shipping facilities, broad swaths of Indian agriculture and textile production were linked to the broader world-economy.

Similar processes took place in Africa and Latin America, as the process of coal-assisted colonialization spread in the late nineteenth century.[20]

As reliance on coal-burning rail and ship systems for military control and commercial trade increased, it became crucially important to ensure that these systems could be supplied with fuel. As a result, British authorities turned their attention to developing a network of coaling stations across the world. Numerous acts of territorial conquest were undertaken by Britain forces to secure strategic sites for these coaling stations. Socotra and Aden were occupied in the mid-nineteenth century, while harbors in Gibraltar, Tahiti, and the Canaries and Cape Verde Islands were developed into major bunkering stations for passing steamers. Port Said, at the entrance to the newly constructed Suez Canal, became the world's largest coaling station, selling over a million tons of coal annually by the late nineteenth century.[21]

Another side of this effort to create a global system of coal-powered transport systems involved increasing government subsidies to private shipping companies. In the mid-1800s, Britain and other European powers began heavily subsidizing merchant marines and mail ships. In exchange for this financial support, commercial companies were required to make ships available for troop transport in times of war. While sailing ships remained important on some transoceanic commercial routes until the turn of the century, by the 1870s tramp steamers had established their dominance over virtually all coastal, river, and transcanal routes.[22]

These forms of state intervention strongly fostered the global expansion of coal consumption. British companies came to dominate this globally integrated network of coal production and trade during the middle decades of the nineteenth century (see Figure 2.2). Bulk shipments of British coal were exported to consumers in the Baltic, France, and as far away as Egypt and Argentina. By 1900 British coal accounted for about 85 percent of internationally traded coal, providing a valuable outward cargo that enabled British shippers to offer competitive home bound freight rates.[23]

Part of the explanation for the commercial success of British firms in foreign coal markets lies in the particular geographic advantages they enjoyed. It was relatively easy to link inland coal mines to coastal port facilities, from which the fuel could be shipped to foreign consumers. Furthermore, dense and smokeless coal from Wales was particularly suited for use in ships, and so demand for this type of coal grew steadily in international markets. Nevertheless, these geographical attributes can only partly explain the dominant position that British coal companies came to hold in foreign coal trade from the mid-nineteenth century to the beginning of the First World War.

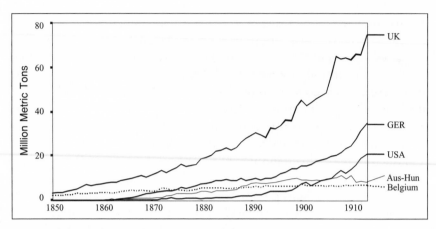

FIGURE 2.2. International Coal Exports, 1850–1913
Sources: See Appendix A.

This early orientation toward export opportunities was also fostered by the political and commercial security offered by the British Empire. As historical research has shown,[24] overseas investments by Continental firms tended to focus on the relatively secure areas dominated by their respective colonial authorities. Dutch investments flowed primarily to the East Indies, French investments concentrated in Algeria and Indochina, and German capital went to parts of Africa and the South Pacific. British investors, however, had a uniquely global terrain on which to move in the nineteenth century. Backed by the strong diplomatic and military protection of their hegemonic state, British entrepreneurs invested heavily in the development of railroads, ports, and other infrastructures all across the world. This allowed for a radical increase in the influx of raw materials and inexpensive foods from Asia, Africa, and Latin America, which spurred further industrial development in Britain.

One additional result of this support from the hegemonic state was an expansion of coal mines throughout Britain's domains. British capital was by far the largest contributor to expanding coal production in colonial areas during the pre–First World War era. From India to South Africa to Australia, investments by British entrepreneurs played central roles in the development of new mines in colonial territories. Small operations were initiated by other colonial powers, of course. French capital was invested in mines in Indochina, for instance, while Dutch entrepreneurs invested in projects in the East Indies. However, British-financed coal production in the periphery far surpassed all other foreign-financed output at the end of the nineteenth century.

The privileged position of Britain in the nineteenth-century world-system also increased the availability of capital that could be reinvested back into domestic mining and manufacturing industries. As a result of a growing influx of capital, a select group of British coal mines went through a remarkable period of expansion during the three decades leading up to the First World War. Although generally still run as family enterprises, British coal firms became very sophisticated organizations. Distinct professions emerged within the mining industry as mine managers, geological engineers, and sales personnel specialized in particular aspects of the maturing industry. In sum, at the turn of the century British coal firms had developed into complex organizations with strong domestic political connections and international investment portfolios.[25]

By the late nineteenth century, a largely British-financed coal system had attained truly global dispersion. Coal-based transportation systems had extended around the world, integrating most colonial regions into European-controlled military and commercial networks. The expansion of coal-powered transportation systems also permitted a rapid growth in the capacity for bulk trading in agricultural and mineral products. For the first time, wide areas of Latin America, Africa, and Asia were securely incorporated into a capitalist world-system that was dominated by Western European countries in general, and Britain in particular. Clearly, the expansions of the modern world-economy and the international coal system were intimately connected processes in the nineteenth century.

The Maturation of the Coal System

The first half of the nineteenth century saw an expansion of coal production and consumption in Western Europe. The second half of the century then witnessed the expansion of this system throughout the colonial world. Interestingly, however, there was also a separate process of diffusion to key semiperipheral regions that was less dominated by colonial elites. In North America, Eastern Europe, and Japan, largely autonomous processes of growth were achieved in coal sectors during the latter part of the 1800s. Due to the emergence of these new areas of production, by the end of the century the coal system had essentially completed its process of global expansion.

The development of a large-scale coal industry in the United States demonstrates again that energy shifts do not simply result from resource scarcities. In the newly industrializing regions of the northeast and central Atlantic states, wood and water resources were still quite abundant when coal mines began to expand in the 1830s and 1840s. Coal also cost significantly more than wood during this early phase of expansion. Still, a small group of entrepreneurs with strong ties to railroad companies and banks played a key role in marketing coal to a new class of consumers on the eastern seaboard in the mid-nineteenth century.[26] Indeed, the evolution of the U.S. coal industry highlights the extent to which market demand for new energy resources can be actively shaped by well-positioned, innovative companies.

In the 1840s, the Lehigh Valley Coal and Navigation Company, the Delaware and Hudson Canal Company, and the Philadelphia and Reading Railroad Company all began promotional campaigns to encourage urban residents and small company owners to switch from wood to coal. Consumers had long resisted using anthracite coal because it is harder to handle and dirtier than wood. To counter this consumer hesitancy, the mine and railroad companies sold stoves, furnaces, and boilers at discounts. Mine owners also urged workers in coal towns to spread the word about the benefits of the fuel (Chandler 1972). By the late 1850s, this small group of companies had spearheaded a significant growth in coal consumption in cities along the Atlantic seaboard.

Coal production was temporarily slowed by the Civil War, but with the end of the conflict the U.S. coal sector began a process of rapid expansion. This growth resulted from a combination of factors. The extensive use of rail transport and metal armaments during the war revealed that modern warfare was becoming increasingly reliant on industrial production. The federal government therefore followed its European counterparts in supporting the expansion of iron, rail, and shipping industries. Meanwhile, new mines in the South and West opened up, and the U.S. coal sector became intensely competitive. These competitive pressures stimulated more rapid processes of mechanization than were typical in Europe, which allowed for greater output from individual mines. Finally, an increasingly large number of migrant miners were drawn into American mines, bringing with them the skills they had learned in long-established mines in Britain, Germany, and Belgium. As a result of these intersecting political, commercial, and social factors, coal production in the United States skyrocketed.[27]

By the end of the nineteenth century, the United States took the lead as the world's largest producer of coal (see Figure 2.3). Most of this coal was consumed in new industrial centers on the East Coast, in growing cities of the Midwest, and in the railroad networks that expanded westward. And, just as coal facilitated European conquest in Asia and Africa, the resource played an important role in the subjugation of Native Americans in the western United States. The massive growth in coal-fired iron and steel

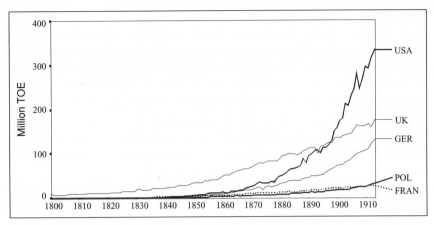

FIGURE 2.3. Five Largest Coal Producers, 1800–1913
Sources: See Appendix A.

industries that occurred in the East allowed a constellation of new tools—from rifles and railroads to barbed wire and plows—to be harnessed in the national project to seize control of the West. By deploying a modern, coal-based industrial system, the United States became the first modern nation-state that was truly continental in scale.

Along with rapid growth in U.S. coal production, mines throughout Eastern Europe also witnessed significant expansion at the end of the nineteenth century. These mines were initially developed by managers from Western Europe. A Welsh entrepreneur, for instance, opened the first mine in what would become a massive coal complex in the Donbass region of Russia and the Ukraine in the 1860s. Capital from France, Prussia, and Belgium also flowed into mines in the present-day countries of Bulgaria, Romania, Serbia, Poland, and the Czech Republic.[28] Indigenous management capabilities, however, grew steadily in the latter part of the nineteenth century. By the time western entrepreneurs and capital left the region during the First World War, Eastern European managers were able to take over operations and increase production on their own accord.

A particularly remarkable case of autonomous development of coal mining capacity can be found in Japan, following its forced opening to the world-system in the 1850s. The Japanese state demonstrated a unique ability to react to the pressures exerted by an expanding, western-centered global system by initiating its own industrialization and militarization programs. Domestic coal mines became a central component of this project of independent development in the late nineteenth century.

Though traditional forms of coal mining had been practiced for centuries in Japan, the Takashima coal mine was the first to incorporate modern, steam-assisted methods. Initially opened by British capital, the mine was taken over by Japanese authorities in the 1870s and underwent a process of expansion that was linked to the nation's broader industrialization drive (Tsutsumi 1998). Virtually all Japanese coal mines were soon absorbed by the industrial combines known as zaibatsu, which also acquired foundries, railways, shipyards, and other major manufacturing and financial sectors. In a process that mirrored German industrialization, the zaibatsu worked closely with the Japanese state to coordinate the production and distribution of resources between sectors. Meanwhile, teams of engineers and managers were sent to North America and Western Europe to learn new skills, while technicians from the West were also hired to work in Japanese mines and factories.[29] As a result of this state-led industrialization program, Japanese coal output grew from around fifty thousand tons in 1870 to eight million tons in 1900, which made it the world's tenth largest producer in that year.

The energy and profits flowing from Japan's coal mines powered an amazing process of industrial transformation. Though the country's first railroad was built in 1872, by 1900 there were more than fifteen hundred miles of rail. Meanwhile, zaibatsu working for the Japanese Navy rapidly developed new ship-building skills, and by the end of the nineteenth century the country was largely self-sufficient in the production of armaments, iron, and steam-powered warships (Uchida 1995). The growing power of this newly industrializing nation was demonstrated when Japan defeated China in 1894 and Russia in 1904, and then seized control of territory in Korea and Manchuria in the first decades of the twentieth century.[30]

In contrast to the remarkable growth of the Japanese coal sector, the Chinese mining sector achieved only modest growth in the pre–First World War period. This is surprising, given the long history of coal mining in China. The industry attracted some investments from both domestic and foreign sources, and a modern mining sector did emerge alongside more traditional mining operations in the late nineteenth century. However, the lack of transportation infrastructure seems to have severely limited coal consumption in many regions before the First World War.[31] In many ways, therefore, China faced the same kinds of hurdles as France, while Japan, like Britain, benefited from the advantages of an accessible coastline. Though Chinese mine operators did amass important technical skills in the nineteenth century, it would take decades before they could contribute large quantities of energy to the country's broader industrialization efforts.

By the beginning of the twentieth century, modern coal mines were in operation on all continents. What was primarily a Europe-centered coal network in 1850 had expanded to become a truly world-scale energy system only fifty years later. Important instances of autonomous coal development occurred in countries like the United States and Japan, while mining in colonial regions also increased steadily. Given all of these coal-related developments, world coal production rose rapidly. In 1800 about 18 million tons of coal were mined, but by 1873 over 300 million tons were being extracted each year. In other words, in the space of only a handful of decades, the world witnessed the maturation of a new global energy system that provided resources on a scale never before witnessed in human history.

Indeed, the period from about 1815 to 1873 represents the first golden age of the modern global energy system. With relative peace and prosperity in Europe, private companies were able to construct the mining and transportation infrastructures required to deliver massive quantities of coal to factories and households. This infusion of coal led to the emergence of many utopic visions and power fantasies in Western Europe and North America in the mid-nineteenth century (Greenberg 1990). Robert Owen and Claude Henri de SaintSimon, for instance, drew up schemes for harnessing coal in ways that they claimed would eliminate menial forms of labor. Even Marx and Engels, in *The Communist Manifesto* (1848), described in awestruck terms the tremendous changes to industry that were being achieved as capitalist enterprises harnessed the power of steam and machines. While they were aware of the dehumanizing aspects of steam-driven factory systems and the appalling loss of life occurring in mines, they still held out hope that the incorporation of new coal resources would set the stage for a transition to a socialist society of material abundance.

Of course, the expansion of the international coal system had contradictory impacts on societies across the world. Coal helped colonial powers tighten their control over vast regions in Latin America, Africa, and Asia, and so international inequalities grew more profound. It powered ships that moved food and raw materials out of colonial regions and into the growing industrial cities of Europe and North America, and in so doing coal intensified an emerging ecological rift between rural and urban areas.[32] While coal fueled the creation of new forms of industrial wealth, it also consumed vast numbers of miners and their families in grueling labor and underground catastrophes. As would become apparent in the coming decades, this energy system brought with it amazing powers of human improvement, along with dangerous new capacities for destruction.

3 Conflict in Coal and the Emergence of New Energy Systems

O ver the course of the nineteenth century, the expansion of the coal system was driven mainly by the interacting dynamics of geopolitical and commercial competition. The industrialization of warfare and colonial conquest spurred state authorities in Western Europe, North America, and certain other regions to support the expansion of coal-related sectors. At the same time, private companies found it profitable to satisfy the demand for coal that was coming from military, industrial, and urban sectors in advanced economies. As a result of these mutually beneficial cycles, the global energy shift toward coal was remarkably rapid and far-reaching.

From the perspective of 1870, it certainly appeared as if coal had a long future ahead of it as the world's primary source of energy. There were ample reserves to be mined in virtually every country, and the demand for new energy resources was strong. Yet, we now know that the international coal system was soon to be challenged by the influx of new resources such as oil. But if there were no problems of resource depletion, what provoked this fundamental and unexpected shift?

 It is my argument that rising social pressures within the coal system played a key role in propelling the shift toward greater reliance on new energy resources. As I will demonstrate in this chapter, labor militancy grew increasingly intense within established coal mining regions in the period leading up to the First World War. As a result, a shift toward new energy resources such as oil and hydroelectricity began to take shape during the late nineteenth century.

Social Conflict in Coal Sectors

The first golden age of capital, and of the global energy system based on coal, lasted from 1815 until 1873. This was a time of peace and prosperity in Europe, and a time during which the deployment of the coal system generated wealth and material advancement for many citizens in the core of the world-system. But this era of expansion came to an end in 1873, when an economic crisis of unprecedented severity swept through Britain and the rest of the industrial world. It was in this period of global contraction that the modern labor movement was forged in Western Europe and North America and began to generate turmoil in centers of coal production.

The Great Depression of 1873–96 was closely linked to the end of the expansionary cycle of the global coal system. In the decades before the Depression, British companies had dominated the construction of rail and shipping facilities in most regions of the world. Tremendous profits were generated as British firms supplied the huge volumes of coal, iron, and other materials required for this mammoth construction boom. By the 1870s, however, companies from France, Prussia, and the United States also began entering the energy and metals markets associated with this infrastructure expansion. As a result, competitive pressures in the world's leading manufacturing sectors began to intensify and profit margins declined.[1] In response to these rising competitive pressures, mining and transport companies across Western Europe and North America attempted to reduce their labor costs by intensifying the work process and cutting wages. It was in this heated context that modern union organizations were formed in coal mines and the first sustained waves of labor unrest occurred in the world's leading mining centers.

We have already seen that the expansion of modern mines in Britain required the transformation of serfs, farmers, and itinerant laborers into a new class of coal miners. In the nineteenth century, similar social transformations took place all across Western Europe and North America.[2] The laborers who were sent to work in these coal mines faced many of the same kinds of coercion that were used in British mines. Like their British counterparts, these workers resisted exploitation as best they could. Indeed, outbreaks of various kinds of labor unrest occurred in mines from the mid-eighteenth century onward, but it took generations before cultural and structural conditions evolved to the point that enduring union organizations could be formed in coal sectors.

As early as the mid-1700s, indentured workers in Britain and Western Europe engaged in work stoppages and destroyed mine machinery in efforts to gain better working conditions and wage improvements. These early efforts by miners to win the right to organize were further stimulated by the transformations wrought in European political culture by the American and French Revolutions.

Indeed, just as the Treaty of Westphalia can be interpreted as the formal inauguration of the systemic dynamics of geopolitical and commercial rivalry, the American and French Revolutions can be seen as granting ideological legitimacy to the systemic dynamic of organized social protest and contestation.[3] A proliferation of new political groups was generated in this new cultural environment, and public demands for increased liberalism, popular democracy, and socialism began to be heard with rising insistence across Western Europe and North America. By the first decades of the nineteenth century, these new currents of radicalism began penetrating mining centers and fostering more determined forms of labor militancy.

Though changes in political culture began to foster the creation of unions in the early 1800s, the social circumstances miners faced in the first decades of the century made it very difficult for them to create labor organizations that could endure attacks from coal owners and state authorities. In the 1830s and 1840s, miners all across Western Europe formed local unions and struck for better wages and working conditions. Coal miners were particularly active participants in the major wave of social unrest that swept through European mines during the revolutionary turmoil of 1848. In virtually all circumstances, though, unruly miners were dismissed by companies and union organizations were dismantled. Fierce repression after 1848 succeeded in destroying Britain's first nationwide coal union, for instance, while local unions in France, Prussia, and Belgium were also driven out of existence.[4]

Beginning in the 1870s, however, coal miners across Western Europe and North America succeeded in forging mass-membership union organizations that were able to survive attacks by owners and political authorities. It was these new organizations that fought against the wage cuts owners tried to impose after the depression of 1873. Indeed, from the 1880s through the end of the twentieth century, coal miners across the world became one of the most militant segments of the working class in the industrialized world.[5]

The key to understanding how this new form of militancy could emerge lies in examining the positional power that came to be held by coal miners in the late nineteenth century.[6] This positional power came from the work conditions in the mines, as well as their location in the wider economies of Western Europe and North America. Within the mines, the skill level required of workers rose steadily during the late nineteenth century. As mine shafts penetrated deep into the earth and machinery became more complex, mining companies became more reliant on skilled workers. Consequently, it became difficult to replace striking miners with new, unskilled laborers. Moreover, the rising costs associated with modern mining meant that any halt in production rapidly translated into heavy financial losses for the mine owners. As countries became ever more reliant on coal resources, these skilled miners found themselves located in an increasingly strategic position within national economies. Strikes in large coal mines could quickly evolve into major commercial crises for an entire country, since so many consumers had come to rely on the resource.

For these reasons, companies found it increasingly hard to fire skilled miners. Even when they did fire workers in particular mines, they were often confronted with sympathy strikes in other mines. Given this new industrial context, coal miners were able to forge enduring union organizations and carry out a first wave of labor militancy in Western Europe and North America that lasted from 1880 until the onset of the First World War.

Just as Britain had seen the first recorded instances of organized labor militancy in coal, it also saw the formation of the first enduring, industry-wide coal union. The Miners' Federation of Great Britain was founded in 1889, just as coal company owners were trying to restore their profit rates by cutting wages in response to the economic contraction of the 1870s and 1880s. The Miners' Federation adopted the new tactic of signing up all kinds of pit workers, from the most to the least skilled workers, and it required all members to contribute a modest sum to their union treasury. As more and more local unions became affiliated with the national federation and became eligible for the federation's strike funds, the success rate of British coal strikes improved. Moreover, it became increasingly hard to isolate the workers in a single pit because sympathy strikes were called when workers in one pit were dismissed from their jobs.[7]

These new union tactics provoked a counterattack in 1893, but efforts by coal operators to dismantle the union were met with the largest strike seen to that point in British history. This protracted and violent coal strike had serious political repercussions. It prompted direct government intervention, and eventually resulted in the passage of national work and safety regulations for the industry. In most pits the miners won substantial wage concessions.

Though unprecedented in its scale and intensity, the coal strike of 1893 proved to be only the opening salvo in an escalating struggle between capital and labor in the British coal industry. Increasingly radical strikes hit the industry in 1910 and 1911, culminating in the national coal strike of 1912. By the first decade of the twentieth century, coordinated union mobilizations in British coal had made this industry the most strike-prone.[8] Repeated strikes not only provoked numerous political crises, but they also resulted in significant wage concessions. They also lent an added edge to the "coal panics" that swept through British society at this time, as citizens became increasingly worried about stability in their most important energy sector.

Similar patterns in the evolution of coal unions can be seen in France. Again, the shift from pit-level unions to wider union federations came in the 1880s. By the middle of this decade, a multipit, mass-union organization had coalesced in French coal that successfully pressed for higher wages. As a result of persistent strike actions, in 1889 unionists succeeded in pressuring mining companies in the largest coal region of France, the Nord/Pas de Calais, to agree to a collective bargaining agreement covering wages and working conditions.

This agreement was a watershed in French industrial relations, as it marked the beginning of institutionalized labor relations in France. With the participation of government representatives in wage negotiations, coal miners were able to win significant concessions from owners. General strikes in 1901 and 1902 and a series of strikes in the decade leading up to the First World War put additional pressure on coal companies and state regulators to improve wages and working conditions.[9] By the onset of the war, French coal miners had not only forged a unique collective bargaining system, but they had also carried out many of the most dramatic strikes in French labor history.

The German mining labor movement was less unified than in Britain or France during the pre–First World War period. By the late 1850s, however, strikes were being carried out in individual mines, and in 1872 more than eighteen thousand miners in numerous pits participated in a six-week stoppage. Labor unrest continued to mount during the next decade, with the strike of 1889 affecting virtually every pit in the Ruhr. The culmination of German labor unrest came with the coal strike of 1905, which was estimated to have been the largest strike witnessed in all of Europe to that date. To ensure that production of a strategic resource wasn't subject to periodic interruption, state authorities forced coal companies to negotiate with unions in the aftermath of each of these large strikes. Significant wage concessions were granted by companies after the strikes.[10]

The same general pattern of union consolidation can be observed in the coal mines of Belgium. By the early 1880s coal miners had emerged as the most strike-prone group of industrial workers in Belgium, though their militant campaigns had rarely resulted in significant wage concessions. This changed, however, with the general strike of 1886, which shut down production in most major coal mines in the country. In the aftermath of this well-organized, lengthy strike, the Belgian government was finally compelled to impose mandatory arbitration on workers and owners in the industry. In 1892 national mining legislation was enacted that set minimum wage and safety standards in Belgian coal mines.[11] In short, after decades of militant campaigns, Belgian coal miners finally succeeded in winning important wage concessions from coal operators.

Across the Atlantic, the evolution of the labor movement in U.S. coal again followed a remarkably similar trajectory. A first period of increased strike activity can be traced to the early 1850s, but for the next two decades strikes were suppressed with increasing violence. It was not until the coal miners joined forces with the Knights of Labor in the 1880s that they were able to form a broad-based federation capable of surviving attacks from public and private agencies. Supported by this national organization, eastern coal miners took advantage of the economic expansion of the 1880s to carry out strikes over wages and working conditions.

In 1890 the eastern and western miners' federations joined to form the United Mine Workers of America (UMWA), marking the emergence of the first national miners' union in U.S. history. The well-organized coal strikes of 1894, 1897, 1900, and 1902 carried out by the UMWA succeeded in forcing coal operators to grant significant wage concessions throughout most of the eastern coal mines. In western mines conflict escalated between coal owners and the Industrial Workers of the World (IWW), culminating in the bloody strike of 1913 in Ludlow, Colorado. By the onset of the First World War, coal had emerged as one of the most strike-prone industries in the United States.[12] By carrying out these militant campaigns, coal miners succeeded in winning large wage concessions in this period.

Similar patterns can also be identified in the Canadian coal industry. In 1877 the first recorded coal miners' union was founded in Vancouver, and the 1880s saw an expansion of unions in other areas. With the assistance of a national federation, Canada's coal miners came to dominate industrial conflict during the period 1900–1913 and succeeded in winning industry-wide wage concessions on the eve of the First World War.[13]

There is a remarkable simultaneity in the timing of union consolidation and spikes in strike activity in Great Britain, France, Germany, Belgium, the United States, and Canada during this period. A major intensification in dynamics of social unrest clearly took place during the late nineteenth century in this group of countries, which together accounted for over 80 percent of the world's coal output. Coal miners were able to push wage rates higher in all of these countries in this era. They were also able to disrupt coal production with some frequency in this period, which heightened commercial wariness. In addition, coal strikes occasionally escalated into major political crises, prompting government intervention in efforts to regulate labor/capital conflict. As this labor pressure intensified, coal's relative position as the world's principal source of commercial energy weakened in relation to new oil and hydroelectric industries that were emerging in the late nineteenth century.

The Emergence of Large-Scale Oil Companies

Just as social unrest was intensifying in the largest coal sectors in the world, new sources of energy began to be tapped by inventors and entrepreneurs. As we will see, important efforts to harness the power of water, natural gas, and solar heat were carried out in the late nineteenth century. However, it was oil that first emerged as a serious challenger to coal's dominance.

Oil has long been used by humans for a variety of tasks. More than five thousand years ago, for instance, people in Sumeria used oil collected from above-ground pools as a medicine, wood sealant, and paint. Oil was then used as a weapon in the siege of Athens in 480 B.C., while Arabian communities later developed the technique of burning purified oil in lamps. In subsequent centuries, citizens in countries all across the world also used small quantities of oil as a lubricant in machines.[14]

Until the mid-nineteenth century, there were two primary sources of commercial quantities of oil. Most of the world's lamp and lubrication oils were provided by whaling fleets that began scouring the oceans in the sixteenth century (Ellis 1991). However, as a wide variety of whales were hunted to the brink of extinction, the price of whale oil began to rise. Efforts to extract oil from shale deposits therefore intensified in England, France, and elsewhere in Western Europe in the mid-nineteenth century. This process was very labor intensive, though, and oil shale operations were unable to satisfy the growing demand for illuminants and lubricants.

While demand for oil was felt most acutely in Western Europe, it was in the United States where the key solution to these oil supply problems was first achieved. A railroad engineer, Edwin Drake, began using salt mining techniques to drill into the ground near an oil seepage outside of Titusville, Pennsylvania. After months of work, Drake's drill hit a pressurized underground pool of oil in August 1859 and the liquid began surging out of the ground. This discovery spurred an oil rush, and within a short time a forest of drilling rigs had sprung up around the area.[15] Simple refineries designed to distill the crude oil into kerosene and lubricants were soon built as well. Meanwhile, the technique of drilling for oil rapidly spread around the world.

The shallow fields of western Pennsylvania allowed a large number of independent wildcatters to begin extracting oil at a low cost. But only a few entrepreneurs had the financial ability to build the storage and transportation facilities needed to get the oil to distant markets. One of these businessmen, John D. Rockefeller, began as a refiner in Cleveland and then diversified into the storage and transport side of the business. Rockefeller and a few others found that by building large tanks near the newly discovered oil fields, they could buy large amounts of the fuel at depressed prices from wildcatters who pumped up more oil than they could handle. To reduce transport costs, Rockefeller also entered into negotiations with regional railroads for preferential rates on oil transport. By providing a steady volume of freight to the railroads, Rockefeller could ease problems of seasonal demand variation that reduced rail profitability. In exchange for this stability, the railroads were willing to offer rebates for Rockefeller oil.[16]

Using these strategies for purchasing oil at bargain prices and transporting it at discount rates, Rockefeller's new Standard Oil Company[17] began a period of rapid growth. This expansion was stimulated by the Civil War, since large quantities of oil were used as lubricant in military hardware and for nighttime illumination. By the mid-1880s Standard Oil controlled around 80 percent of U.S. refining capacity, and by 1910 the company accounted for approximately 90 percent of all refined oil products sold in the country. Standard Oil also turned to international markets, coming to control over half of the world's sales of refined oil by 1900.[18] However, Standard Oil's position began to be challenged by foreign and domestic competitors around the turn of the century. These new companies in many cases replicated the competitive strategies that had been pioneered by Rockefeller.

Just as Standard Oil had grown out of the oil boom in Pennsylvania in the 1860s, a series of major companies were founded in the midst of a larger oil boom that took place in Russia in the 1870s. The Nobel brothers, for instance, succeeded in constructing a series of pipe and rail lines to move oil from Baku, located on the land-locked Caspian Sea, to Batum, a city on the Black Sea. They followed these engineering achievements with yet another—that of converting ships into tankers for the transport of their oil from the Black Sea to the Mediterranean. With their refining and transport systems in place, the Nobel brothers were able to use low-cost Russian oil to compete against the Standard Oil Company in Western Europe in the 1880s.[19]

The Nobels also entered into a venture with a British citizen, Marcus Samuel, to set up a company to market Russian oil in the Far East. The key to this effort was gaining permission to ship oil through the Suez Canal. The British government, which operated the Canal, had denied permission to Standard Oil to move oil along this route because of concerns about the danger of explosions. Samuel arranged for the construction of new oil tankers that met more stringent safety standards, and then began marshaling his contacts in the British government to his cause. Samuel received permission in 1893 to move oil through the Canal. Samuel's new firm, the Shell Transport and Trading Company, was soon selling large quantities of low-cost Russian oil throughout the Far East.[20]

Another major private competitor emerged during this period as well. The Netherlands Royal Dutch Company, founded in 1890 by Henri Deterding, initially drew its oil from fields in Sumatra. Within a decade Royal Dutch had built marketing and distributing facilities that stretched throughout China, Malaysia, and India. Deterding soon entered into an alliance with Samuel's Shell Oil to counter aggressive attacks from Standard Oil, and in 1907 the two companies merged. The Royal Dutch Shell enterprise proved to be a strong competitor in world oil markets. From its initial holdings of oil in Russia and the East Indies, Royal Dutch Shell expanded into Texas, California, Mexico, and Venezuela in the early twentieth century.

Each of these private oil companies had tense relationships with state officials before the First World War. In the United States, Standard Oil was treated with hostility by state and federal authorities. For their part, the Nobels received little direct assistance from any government. Meanwhile, apart from being granted permission to move oil through the Suez Canal, the Shell Oil Company received virtually no support from the British government until the First World War.[21] And though Royal Dutch was initially chartered by the Dutch government, it remained almost entirely self-financing and independent for the first two decades of its existence. Most governments were focused on supporting their domestic coal industries at this time rather than on fostering the growth of a new energy industry.

The growth of these major foreign competitors steadily eroded Standard Oil's dominant position in the international oil industry. Serious challenges also emerged in the domestic U.S. industry, especially as new discoveries of oil shifted the center of the industry away from Standard Oil's strongholds in the East. The massive oil discovery in 1901 at Spindletop, Texas, for instance, generated two new U.S.-based rivals of the Standard Oil Trust. One of the companies, Gulf Oil, received substantial backing from the Mellon Bank, which had started with extensive investments in the Pennsylvanian coal industry (Sampson 1975). With this financial support, Gulf developed a network of tankers to deliver cheap Texas oil to customers throughout the United States and Europe. The other company, Texaco, built terminals on the Gulf of Mexico and began shipping oil to U.S. and Latin American ports. Meanwhile, additional discoveries of oil in Oklahoma and California allowed for the expansion of other independents, including Union Oil, Atlantic Refining, and the Vacuum Oil Company.

In the midst of this rising competition from foreign and domestic rivals, Standard Oil was confronted with yet another challenge. After years of delving into the company's secretive operations, U.S. authorities came to the conclusion that the Standard Oil Trust had engaged in restraint of trade in the domestic oil industry. In 1911, a federal judge announced that the Trust had six months in which to dissolve itself into its constituent parts.

In hindsight it is generally agreed that this dissolution helped the Standard Oil conglomerate break out of a period of organizational malaise. For at least a decade prior to the court action, the Trust had been hamstrung by an aging executive committee that attempted to run too many facets of the company's business (Chandler 1962). The dissolution forced the directors of each of the constituent companies to take more independent control of their operations, which increased their competitive behavior. Some of the new companies—especially Standard Oil of New Jersey, Standard of New York (Socony), and Standard of California (Socal)—threw themselves into the quest for new oil fields and markets with renewed aggressiveness after the dissolution.

By the eve of the First World War, a select group of private oil companies had achieved important commercial milestones. American and European oil companies had built networks of oil wells, pipelines, and tankers that stretched throughout the world, and the wealth generated by these ventures was astonishing. In fact, a survey of the world's 100 largest firms carried out in 1912 revealed that Standard Oil of New Jersey was second only to U.S. Steel in terms of its declared market assets.[22] Royal Dutch Shell, Standard Oil of Indiana, Socony, and Socal were all listed in the top 50, while the world's largest coal company, Pittsburgh Coal, was listed as number 81. Even though oil companies had only completed a first phase of growth into illuminant and lubricant markets, they had already amassed financial resources that far outstripped the world's largest coal companies.

One reason that these oil companies were able to accumulate wealth so rapidly was their ability to contain labor militancy. Whereas large numbers of coal miners came to occupy strategic positions within centralized mining operations, the occupational characteristics of oil served to dampen labor militancy. Oil production involves a wide variety of distinct tasks, including drilling, pipeline construction, well maintenance, transportation, and refining. Each of these categories of work requires specific kinds of laborers and tends to result in distinct modes of labor control that reduce the capacity of workers to create unions and engage in strike activity.

For instance, the workers who drill wells and construct derricks might be expected to form radical unions, given the physically demanding and hazardous nature of drilling work.[23] The task of forcing segments of pipe deep underground with hydraulic drills is strenuous, and workers must contend with a constant threat of fires and explosions. The fact that drilling is frequently carried out in inhospitable deserts, jungles, and tundras or on ocean platforms heightens the challenges faced by workers at this stage of oil production. Drillers are also generally the most skilled laborers to be found out in the oil fields, and they enjoy a certain degree of autonomy on their rigs. In many ways, therefore, they resemble coal miners who work in underground shafts.

However, drilling crews have the advantage of working in an industry whose profits derive primarily from capital-intensive, mass-production systems. For instance, if in the early decades of the twentieth century labor costs in U.S. coal represented over 70 percent of total production costs, during the same period the proportion of labor costs in oil production was estimated to be under 10 percent.[24] Most oil companies have therefore been able to pay their drilling crews relatively high wages without significantly reducing their profit rates. This feature of oil production provides one explanation for the generally low level of overt labor conflict found in most regions of the world over the history of the industry.

In addition to the skilled drillers, oil fields and pipeline systems are serviced by an assortment of mechanics, welders, heavy-equipment operators, and other workers. The number of laborers on any given field fluctuates widely, however. Large numbers employed during construction phases generally give way to small maintenance crews once oil derricks and pipelines have been installed. The transitory nature of most oil field employment provides another explanation for low levels of militancy, since it has proven difficult to organize and sustain unions in such fluid contexts (Davidson 1988). The employment of subcontracted or expatriate workers in semiskilled oil field occupations, a strategy carried the furthest in the Persian Gulf, has further hindered the development of enduring labor organizations.

In contrast to drillers and oil field laborers, workers employed in oil refineries have come to occupy strategic positions within oil industries. The refining process involves the application of heat, pressure, and chemical agents to crude oil to extract impurities and produce standardized categories of fuel. In the early years of the industry the refining process was relatively simple, but by the first decades of the twentieth century the development of high-performance engines requiring very specific types of fuel led to the establishment of sophisticated, and expensive, refinery complexes. The chemists, engineers, machinists, electricians, and safety personnel employed in these installations have generally been quite skilled, and their labor turnover rates have been low.[25] With the expansion of refineries, therefore, large concentrations of skilled workers have come to occupy particularly advantageous positions within the economies of specific oil exporting and consuming countries. To the extent that labor militancy has emerged within oil sectors in specific countries, it has generally been focused on these refinery operations.

Labor relations in oil industries have, therefore, generally been more tranquil than those in coal. Of course, we will see that there have been important strike waves that have rippled through this energy system. However, in the first half of the twentieth century oil companies did not have to contend with sustained unrest. What companies did have to find, though, was a key technological application that would bring oil into the center of a new system of industrial production and consumption. In other words, oil companies needed an analog to the steam engine that had driven the coal system to global dominance.

The Early Development of Oil-Based Technologies

In its first phase of development, the oil industry provided products that supported the preexisting coal system. Kerosene lamps were used in coal mining, for instance, while oil-based lubricants were consumed in coal-powered rail locomotives and ships. In fact, without the ability to switch from whale to mineral oil, coal-based systems would have run into severe difficulties. In the late nineteenth century, however, oil shifted from being a complement to coal to becoming a direct competitor. This shift occurred because of the development of the internal combustion engine, which brought oil into commercial rivalry with coal in key transportation sectors across the world.

Just as in the case of the earlier development of the steam engine, the invention and diffusion of the oil-powered internal combustion engine was not a natural or automatic process, and it was certainly not a technological fix for pressing material constraints. At the time the internal combustion engine was developed, there was little need for additional energy resources in Western Europe or North America. Coal was abundant, and steam-based systems were able to meet established energy needs. In the crucial transportation sector, for instance, steam-powered trains and ships showed no limit to their capacity to move huge quantities of goods and people. Steam engines were even being introduced into personal cars and tractors in the late nineteenth century, and companies were beginning to market these steam vehicles in Western Europe and North America.

In the United States, for instance, the Stanley Brothers manufactured steam cars that functioned better than gasoline-driven cars in the first decade of the twentieth century. The Stanley Steamer was not only faster than gas cars of the time, but it was a much simpler vehicle to operate. The power of the boilers could be easily adjusted, whereas gasoline-powered vehicles required complicated carburetor, clutch, and transmission systems to adjust the thrust of their engines. Moreover, the widespread availability of coal provided what seemed to be an insurmountable advantage to steam cars. Indeed, analysts at the turn of the century argued that gas autos would never enter into widespread use since there were few petrol stations. Many also argued that oil-based engines should not be allowed onto city streets, since they were thought to be inherently prone to explode.[26] The oil-powered internal combustion engine therefore seemed to face insurmountable odds in the marketplace. And yet, only a few decades after its invention, the internal combustion engine had established almost complete dominance over key transportation sectors. This revolutionary shift highlights again how rapidly change can occur in the world's energy sectors.

The invention and early adoption of the internal combustion engine resulted from a particular constellation of commercial and social factors. The world-economy entered a new phase of growth after 1893, for instance, which freed up capital for experiments on new energy technologies and machines of all kinds. This general economic expansion also generated a class of affluent consumers in Europe who were able to purchase experimental contraptions such as bicycles and, eventually, automobiles. Meanwhile, the spread of commercial patent laws across Western Europe after 1870 opened up the possibility that new inventions could turn a profit for private individuals—even ones who were not well connected to an aristocratic elite.[27] Finally, the proliferation of world fairs and international exhibitions in the late nineteenth century provided a social context in which medium-sized companies could demonstrate their new inventions to the media and potential customers.

The first inventor to devise a gas-powered internal combustion engine for commercial sale was the Frenchman Jean Lenoir. This engine was not particularly efficient, however, and only around five hundred of the Lenoir engines were ever sold.[28] The German team of Nicolaus Otto and Eugen Langen then created a four-stroke engine with better energy efficiency. Their company debuted the new four-stroke engine at the 1867 Paris Exposition, where it was declared to be the most successful auto engine of the day. After their successful demonstration at the exposition, they received a large number of orders for their engine. Their company, in fact, became the first to mass-produce gasoline-powered automobiles for commercial sale.

In 1872, the German engineer Gottlieb Daimler joined the Otto-Langen company as production manager. Daimler had worked on steam engines in France and England and was intimately aware of the most advanced steam technologies. Under his direction, the Otto-Langen company continued to improve the efficiency and durability of the gasoline engines. Their best engines were again successfully demonstrated at numerous European expositions in the 1870s and 1880s. Companies from all over Europe and the United States purchased rights to manufacture the Otto-Langen engines. For a time, the company had protection from a broad patent and they dominated internal combustion engine production. In 1884, however, their patent was overturned and a new phase of competitive development ensued in the emerging automobile industry.

One source of competition for the Otto-Langen Company came from Daimler himself, who left the company to pursue his own fortunes in the early 1880s. In 1886 the Daimler Company debuted what was then the most powerful and efficient gasoline engine, which it aggressively promoted through subsidiaries in Germany, England, France, and the United States. At the same time, yet another German entrepreneur—Karl Benz—began mass-producing his own gasoline-powered engine. Only a few years later, the German engineer Rudolf Diesel emerged with a compression-based internal combustion engine that was even more efficient and powerful than spark-ignited gasoline engines. In the 1880s, in short, a series of German companies made rapid progress in improving the performance and efficiency of various kinds of internal combustion engines.

Although German inventors were the technical pioneers, it was in France where autos first entered into widespread use. Gasoline-powered cars were at first purchased by the aristocratic elite, and they were mainly used in auto races. The world's first auto race, in 1894, was actually won by a coal-powered tractor. But second and third places went to gasoline cars, and by the following year gasoline-powered cars had come to dominate French auto racing. These auto races played an important role in the development of the gasoline-powered car, as they provided the perfect context in which to experiment with new configurations of engines, chassis, and other associated components. They also enhanced the mystique of the new contraptions in France and throughout Europe.

A group of French auto companies, which regularly ran cars in the races, made important contributions to the design of gasoline-powered vehicles in the latter part of the nineteenth century. The French company Michelin, for instance, developed the pneumatic tire. The Panhard-Levassor company then introduced the rear-wheel-drive train. By putting the engine in the front of the car and using a drive shaft to propel the vehicle from its rear wheels, the company was able to put larger engines in their cars while simultaneously improving the maneuverability of the vehicles. The Peugot and Renault companies, meanwhile, made important improvements in car-buretors and fuel delivery systems (Laux 1963). As a result of these inno-vations, France led the world in petrol-powered auto manufacturing at the turn of the century.

European companies also led the world in introducing internal combus-tion engines into marine and rail systems. In 1903 Rudolf Diesel installed a small diesel engine into a canal boat, while the next year an engineer in charge of the marine division of the Nobel Brother's firm installed a diesel into a small oil tanker. In 1912 the Dutch vessel Selandia became the first ocean-going passenger ship to be powered by a diesel engine. In that same year, a German company built the first diesel-powered locomotive.[29]

Even in the arena of aircraft propulsion, European firms took an early lead away from American inventors. Wilbur and Orville Wright had used a small American engine to power their airplane in their inaugural flight in 1903. By 1908, the Wright brothers were touring Europe to demonstrate their machines, with the goal of generating sales for their new company. However, European auto and manufacturing firms soon began producing their own planes. Indeed, it is estimated that between 1911 and 1913, over one hundred European companies were created to build and sell airplanes. By the eve of the First World War, European companies were manufacturing many more airplanes than their American competitors.[30]

At the beginning of the twentieth century, it was clear that oil-powered engines could deliver performance capabilities that were comparable to coal-based engines, but the gasoline-powered engine still faced signifi-cant hurdles to its wide adoption. There was a very limited infrastructure for delivering gasoline, and concerns about explosions were widespread. Moreover, the gasoline car was mainly marketed as a luxury product to an elite clientele. Entrepreneurs across the Atlantic, however, soon devised new strategies for manufacturing and selling gasoline-powered vehicles that revolutionized automobile production—and the oil industry that pro-vided their fuel.

The Emergence of an Oil-Based System in the United States

Entrepreneurs in the United States initially lagged behind Europe in the development and marketing of vehicles powered by the internal combustion engine. By the 1910s, however, the United States emerged as the leader in the production and marketing of vehicles run on oil. As a result of successes in the mass diffusion of gasoline-powered vehicles, it was the United States that witnessed the first full consolidation of an oil-based energy system.

To explain this unexpected turn of events, we have to again turn to an analysis of the convergence of circumstances that fostered this outcome. Just as a constellation of material and social factors fostered the emergence of a coal-based system in Britain in the nineteenth century, a new set of conditions intersected in the United States to produce the emergence of an oil-based system there in the twentieth century. Again we will see that commercial and social factors were far more important than geographical factors in producing this outcome. Also, an analysis of the expansion of the oil system will again demonstrate how new energy technologies can grow and thrive, even in a context where established energy systems are meeting the power requirements of a society.

Let us begin by examining the material context that existed at the time an oil-based system emerged in the United States. On the one hand, a strongly supportive factor was provided by the discovery of massive new oil reserves in Texas and other western states soon after the turn of the century. However, distribution systems for oil were scarce in most regions of the country. Therefore, while there was a new abundance of oil resources in the western United States, there was only a limited capacity to get these resources to key markets in the East. Indeed, promoters of steam-powered cars based their arguments in large part on the fact that distribution systems already existed for coal in most regions of the country. Given that material factors largely canceled each other out in the United States, we need to turn to an analysis of commercial and social factors to understand why oil-powered cars began a rapid process of diffusion in this country rather than in Europe.

It has already been noted that French and German companies marketed their cars to an elite segment of consumers. This was largely due to the fact that the vehicles themselves were hand-crafted by skilled artisans, which resulted in high production costs. European companies tried to scale up their operations, but mainly by hiring new craftsmen to build each machine from the ground up. As a result, per-unit costs did not decline appreciably. A similar manufacturing style was followed by steam engine companies, such as the Stanley Brothers Steam Motor Company, and so the production costs of these machines also remained high.[31] With these high manufacturing expenses, most auto manufacturers were forced to target their sales to an affluent class of consumers.

In the region around Detroit, however, a fusion of new styles of production and marketing occurred that allowed cars to be sold to a much broader class of consumers. Though Henry Ford has been credited with designing the new system of auto manufacturing that emerged in Detroit, it is more accurate to see the innovations that transformed the American auto industry as emerging from a host of incremental changes in the broader U.S. economy and society. In many ways, these changes were linked to the fact that the United States was a resource-rich country that was coming to occupy an increasingly central position in the world-system at the beginning of the twentieth century.

One change of central importance was the emergence of what became known as the American style of manufacturing (Hounshell 1984). This approach involved the mass production of identical items in a factory setting. Wherever possible, the tasks to be carried out in the factory were divided into separate, simple jobs, which allowed low-skilled workers to be employed. By putting any able-bodied person to work doing a single, repetitive task, it was demonstrated that quite complex items could be manufactured in large quantities.

This style of production was first employed in the Springfield Armory in the 1790s. The mass production of armaments was then radically expanded during the Civil War. By the end of the nineteenth century, the mass-production assembly line was being used to produce farming equipment, bicycles, and meat products in the United States. These techniques were eventually adopted by the Oldsmobile company, and then by Henry Ford and other car companies in the Detroit area. A burgeoning population of new migrants from Europe helped fuel the growth of these new companies, while a growing abundance of energy and metals resources in the Great Lakes region further supported the expansion of the automobile industry.[32]

In addition to pioneering new production techniques, U.S. auto companies also began to make use of a new, uniquely American system of mass marketing. As Chandler (1962) has demonstrated, the expansion of railroads across the United States led to the emergence of companies with separate sales and catalog divisions dedicated to promoting products on a national level. Many companies also began opening franchises in city after city to sell their products to a broader base of consumers. Chicago became an important center for these new mass-marketing companies, and these sales strategies also spread to the neighboring Detroit area in the late nineteenth century.

Although he did not independently invent these new styles of manufacturing or marketing, Henry Ford was the first to integrate them into a single strategy of automobile production and promotion. The employment of new mass-production techniques allowed Ford to radically increase the number of cars produced, while reducing the cost on each unit. By 1914 the basic Ford car had declined to a price of $500, which was within the budget of a wide range of urban and rural citizens (Barker 1985). Ford also opened dealerships in cities all across the country that offered credit, replacement parts, and licensed mechanics to new consumers. Later entrepreneurs such as Alfred Sloan of General Motors refined these mass-production and mass-marketing systems to allow for the production of a wider variety of models and styles of automobiles. As a result of these new manufacturing and marketing strategies, sales of gasoline-powered vehicles skyrocketed in the United States.

In contrast to the new dynamism of the American gasoline-based auto industry, the production and sales strategies used by steam-engine companies remained largely unchanged. The Stanley Steamer Company, for instance, steadfastly refused to shift to mass-production techniques. The company also failed to open dealerships outside of their original markets in the Northeast (McLaughlin 1954). The failure of the steamer to become the standard form of automobile was therefore as much a product of entrepreneurial stagnation as technical drawbacks.

Meanwhile, European auto companies generally remained wedded to artisanal modes of production until after the First World War. In many ways, this resistance to change was a product of early successes achieved by artisanal production systems. By selling a smaller number of hand-crafted vehicles to an exclusive clientele, the European companies were able to earn high profits on each unit.[33] In most respects, the vehicles that were produced in Europe were of a higher quality than those manufactured by U.S. assembly lines. It was only after U.S. mass-production systems had come to dominate the world's automobile markets that European companies were forced to revise their own manufacturing and marketing strategies.

With the utilization of new mass-production and mass-marketing techniques, U.S. auto production quickly outstripped European production. Only around six hundred motor vehicles were produced in the United States in 1900, compared with over one thousand in France that year. By 1907, however, U.S. annual production of motor vehicles had soared to forty-three thousand, while French production stood at twenty-five thousand. The U.S. lead grew even wider after the First World War, reaching an annual production rate of over 3.5 million vehicles in 1924, compared to a French production rate of 145,000, English production of 133,000, and German production of only 18,000.[34]

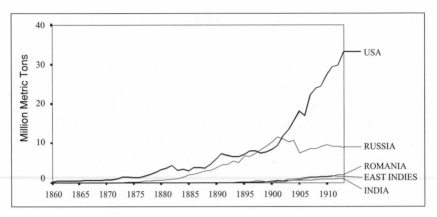

FIGURE 3.1. Five Largest Oil Producers, 1860–1913
Sources: See Appendix A.

The different approaches to the production of gasoline-powered vehicles found in the United States and Europe are also illustrated in the
emergent airline industry. Just before the First World War, companies
in the United States began experimenting with the mass production of
standardized aircraft, but these airplanes proved to be of very low quality.
Indeed, American airplanes were shunned by European pilots during the
war. In contrast, European companies again relied on highly trained craftsmen to produce higher-quality airplanes. Therefore, in its initial stages the
artisanal approach to manufacturing airplanes appeared to be superior to
the mass-production style. By the 1930s, however, American manufacturers had improved their mass-production operations to the point that they
could turn out large numbers of reliable airplanes.[35] From that point on,
European companies had to struggle to make the transition to American
styles of manufacturing.

Given this rapid growth in the sales of vehicles and airplanes, demand for
gasoline and lubricants surged. This rising desire for oil-derived products
was then met by a rapid growth in oil production. In 1860, around one
hundred thousand tons of oil was produced by the world's oil wells. Only
fifty years later, wells were providing over 52 million tons per year, which
represents a more than 500 percent increase. As Figure 3.1 shows, the
United States outstripped all other countries in oil production after 1900,
with wells in Russia, Romania, the East Indies, and India providing most
of the rest. Clearly, this oil system was of international scope from its very
inception.

In examining the consolidation of this oil-based system, we have again
witnessed key features of most modern energy shifts. The shift to greater

reliance on oil was not driven by resource pressures. On the contrary, the rise of the oil system occurred just as coal was entering into full maturity. Instead, growing demand for the new resource was generated by commercial and social transformations. Just as specific financial and social conditions had fostered the mass deployment of the coal-powered steam engine in Britain in the nineteenth century, new commercial and social conditions favored the rapid diffusion of the internal combustion engine in the United States a century later. The successful development of an American style of mass production and marketing jump-started what otherwise would have been a slow trajectory of growth in demand for oil. Given these uniquely favorable societal factors, the international oil system began a process of rapid growth that would only accelerate in the coming decades.

The Early Development of Electrical Systems

In addition to the expansion of oil, the late nineteenth century witnessed another important transformation in the global energy system: the creation and rapid expansion of electrical systems. The development of these systems brought a whole new realm of energy services to businesses, communities, and governments in the industrial world before World War I.

With the rise of the electrical industry, a new level of complexity was introduced into the global energy system. As noted in Chapter 1, electricity does not occur in nature in a form that humans can make direct use of, but other naturally occurring forms of energy—like water, coal, and natural gas—can be transformed into electricity. This electricity can then be transmitted to locations where it can be consumed in lights or electric motors. The advent of new electrical systems therefore greatly increased the flexibility of existing energy systems, while at the same time it began to obscure the true material bases of energy production and consumption.

The rise of the electrical industry has certain parallels with that of coal and oil, though it also differs in important ways. As with both coal and oil technologies, the development and initial expansion of new electrical systems was not driven by resource constraints. Instead, the growth of electrical networks reflected the growth in scientific research and commercial experimentation that accompanied the global economic boom of the 1893–1913 period. The expansion of electrical systems also responded to demands by commercial enterprises for more flexible power in the workplace and to the desire of affluent consumers to have cleaner sources of illumination and power in their homes.

The countries with most success in early electricity development—most notably the United States and Germany—were nations with strong scientific and technical infrastructures, supportive business environments, and social contexts that favored the adoption of new energy systems. The pioneering role of U.S. electricity entrepreneurs was fostered in particular by the emergence of New York City as a key world financial center, which generated a sizable pool of surplus capital that could be invested in the new energy technologies. Meanwhile, in Germany the development of a strong system of technical institutes fostered early innovations. Conversely, in Britain an extensive system of coal gas lighting utilities slowed the expansion of electrical grids.

Electrical systems also supported the continued expansion of the pre-existing coal system, since a sizable share of the electricity consumed was produced in coal-fired engines and turbines. Again we witness the fact that newer energy systems often begin their history as allies to more established energy regimes. And, again reflecting the pattern observed in oil, technological breakthroughs in electrical systems were first made by private entrepreneurs, though the scaling-up of these systems required extensive support from governmental agencies.

A key difference in emerging electrical sectors, however, was the highly centralized nature of the industries. In fact, electricity became even more dominated by a small group of companies than oil. Four firms—General Electric, Westinghouse, Siemens, and the Allgemeine Electrizitats-Gesellschaft (AEG—Germany General Electric)—came to control the international electrical industry from a very early period. Even in contexts where electrical systems were run as public utilities, these public trusts were still highly dependent on these four firms for access to electrical components and expertise. We can therefore discern something of a historical pattern, in which the corporate organization of energy systems shifted from one of intense competition (coal) to relatively strong oligopoly (oil) to regulated monopolies (electricity). This pattern raises the question as to whether a future energy system, based on hydrogen, for instance, will follow the trajectory toward further ownership concentration. While it is too early to tell how a new system will be organized, there is no inherent reason to expect a deviation from this pattern of increasing centralization of corporate control in new energy systems.

In turning our attention to the early innovations in electrical systems, we also witness another pattern that resembles the developmental trajectories of coal- and oil-based technologies. The early research and development of electrical systems emerged from a context in which tinkerers, scientists, and entrepreneurs from across the developed world exchanged information and built on one another's experiments.

The English researcher William Gilbert was apparently the first to systematically study the phenomenon of static electricity, and it was he who coined the term "electrification" around 1600. Further investigations into static electricity were then conducted over the next century by European and North American scientists. Benjamin Franklin demonstrated that lightning was a larger version of static energy in the mid-1700s, for instance. In the late 1700s, the Italian researchers Galvani and Volta conducted pioneering work on electrochemical reactions and batteries. In the first decade of the 1800s, a variety of European scientists turned their attention to examining the magnetic field that surround electrified wires.

One of the key breakthroughs in the history of electrical research was achieved by the English scientist Michael Faraday in the 1820s. Interestingly, just as with earlier coal and oil systems, Faraday's research was conducted with support from the British government. In 1813 Faraday became an assistant researcher at the Royal Institution in London, where he worked for the next two decades. It was in this supportive context that Faraday uncovered the phenomenon of electromagnetic induction and devised the first electrical motor. French, German, Belgian, and Swiss engineers then made improvements to Faraday's dynamos and motors.[36]

By the 1830s, key electrical technologies had been demonstrated in laboratory contexts. However, these systems remained confined to experimental workshops for the next fifty years, largely because capital investments in energy systems were focused on coal- and then oil-based technologies. The global economic upswing of the late nineteenth century, though, generated new pools of venture capital and provided the opportunity for already existing electrical technologies to be perfected and installed on a large scale.

The individual who has received most credit for paving the way for the large-scale commercialization of electrical systems is Thomas Edison. Indeed, Edison and the researchers gathered in his laboratory at Menlo Park set up a methodological system of addressing technical and commercial problems that resulted in the development of reliable electrical products and interconnected systems. For instance, the Edison group invented a more durable filament for light bulbs, and they also worked on methods to subdivide currents into particular units. In 1881 and 1882 they inaugurated the world's first comprehensive electrical systems in London and New York City. This research group, which eventually took on the name of General Electric, established itself as a leader in the development and commercialization of 110-volt, direct current (dc) electrical components and systems.

Although Edison's General Electric Company was often at the forefront of electrical developments, it was quickly challenged by a handful of other companies as well. The most persistent competition initially came from the Westinghouse Company. George Westinghouse emulated Edison in creating an industrial research laboratory in which new appliances, transformers, and transmission equipment were developed. Westinghouse decided to focus on the higher-voltage alternating current (ac) system, however, instead of Edison's lower-voltage dc system. For a time, the Edison and Westinghouse companies engaged in intense competition to ensure that their specific systems became the industry standard. This well-documented "battle of the systems" generated a whole set of important technological innovations.[37] New transformers, polyphase motors, and electric gadgets began to flow from the laboratories of the two competing firms on a regular basis at the turn of the century.

Technological developments in the electrical sector were also stimulated by a similar competitive dynamic that arose in Germany. Just as in the United States, in Germany two companies came to dominate that country's electricity industry. The first, the Siemens and Halske firm, initially specialized in designing and installing telegraph equipment for German municipalities and military operations. Indeed, this company had particularly strong connections to officials in the Prussian and then German governments. Werner Siemens had received extensive training in military academies, and had then spent 15 years in military service. After having developed a particularly successful electric motor, he was therefore well positioned to shift his company's focus to providing lighting and motor power for municipalities across Germany.[38]

The second German electric company to rise to global prominence had very different origins. In sharp contrast to the Siemens Company, the Allgemeine Electrizitats Gesellschaft (AEG) was cofounded by two entrepreneurs who had few contacts with the German state. Emil Rathenau and Oskar von Miller instead had to rely entirely on private investments to finance the construction of long-distance electrical transmission lines, electric trolley car systems, and Berlin's first electrical system (Bowen 1950). They were quite successful in these schemes, and by the turn of the century AEG had emerged as one of the world's four largest electrical manufacturing firms.

Taken together, the entrepreneurial successes of General Electric, Westinghouse, Siemens, and AEG were nothing short of remarkable. In a context of burgeoning supplies of coal and oil, these companies were nevertheless able to create markets for a wide range of new electrical products and services in many countries across the world. Interestingly, the Siemens

Company was the first electrical firm to take its operations international. Already in the 1850s the company had begun building telegraph systems in England, Russia, and Austria, while in 1879 Siemens put China's first electrical generator into operation. But the other electric companies soon began to expand internationally as well. By the turn of the century, subsidiaries of the four leading electrical firms were selling products throughout Europe, Latin America, and Asia. Each of these companies emulated the earlier example of automobile manufacturers by making use of international technology exhibitions to publicize their new systems to regular citizens, captains of industry, and government officials. As a result, by the 1890s a range of consumers in many parts of the world were purchasing new electrical systems.[39]

In all of these countries, electricity made its entrance first as a source of light for commercial and industrial centers. Then, with the development of new electrical motors, manufacturing companies began installing machines that were powered by electricity rather than coal-driven shafts. The electrification of the factory in many ways permitted the development of the continuous flow assembly line, since electric power could be delivered to specific machines much more efficiently and precisely than bulky and dangerous systems of shafts and chains (Nye 1998). In an important way, therefore, electrical systems fostered the Fordist mass-production mode that was in turn generating growing numbers of oil-powered autos and machines.

Electrical power was not only introduced into factories, but it also came to be used extensively in urban transit industries. Although transportation sectors would become dominated by gasoline-powered vehicles in the mid-twentieth century, there was an earlier period in which electric-driven trolleys, trains, and buses grew to dominate some city landscapes. Again, the United States led the way in adopting electrified transit systems. At the turn of the century, while cars were just beginning to appear on city streets, U.S. citizens were taking almost five billion trolley rides a year. Cities across Western Europe also witnessed a dramatic expansion of electric-powered mass transit systems in the period 1880–1913. Even in some non-Western cities, such as Buenos Aires, Johannesburg, and Tokyo, electric trolleys were built along major urban streets before the First World War.[40] In most cases, cities witnessed the emergence of what we might call electrical growth machines, consisting of land owners, real estate developers, construction unions, and local political bosses. Together, each group saw a benefit in constructing electrical grids servicing manufacturing centers, retail stores, and private homes.

Taken together, the construction of central power stations, factory motor systems, and networks of street lights and trolley cars reflected a wave of capital investments that matched the earlier boom in steam-powered railway construction—and actually outstripped investments directed toward oil-powered systems. As with earlier infrastructure investment booms, state agencies across the world played an important role in stimulating and managing these massive electricity investments. Throughout Western Europe and Japan, federal and municipal authorities were intimately involved in raising capital and setting prices for electricity services. Even in the United States, where private capital played a more central role in the development of street light and trolley systems, government authorities carefully regulated prices and service conditions in the new electricity sectors.[41]

In addition to the supportive interventions of government and entrepreneurial groups, electrical system expansion was fostered by favorable labor conditions. In most countries, there were few strikes carried out by workers employed in generating plants, grid construction, or line maintenance activities. In the United States, for instance, the National Brotherhood of Electrical Workers successfully pushed for higher wages without having to engage in work stoppages. This was partly because the increasingly high-tech nature of electricity generation allowed more power to be produced with proportionately fewer workers. Relative labor tranquility was also achieved because utilities generally constructed grids with multiple sources of primary power, so that a strike in one or even several generating plants did not affect a broad number of customers.[42] While coal sectors were being increasingly rocked by strikes in the pre–First World War period, electricity industries were characterized by labor peace. This factor was important in stimulating shifts away from coal gas-powered lights and rail transit in urban areas across North American and Western Europe.

Given this growing demand from industrial, residential, and transportation sectors, electric power generation began a rapid process of growth. Only thirty years after the first electric generator went into operation in the United States, over 24 billion kilowatt hours (kWhs) were being consumed on an annual basis. The runner-up was Germany, where around five billion kWhs were consumed on an annual basis, while the United Kingdom followed in third place with around two billion kWhs consumed on the eve of the war. Italy, Switzerland, France, and Norway also each consumed over one billion kWhs per year at the time. Interestingly, only thirty years after electricity became a commercial product, it was being consumed in large quantities in areas outside of Western Europe and North America as well. In Japan, for instance, over six hundred million kWhs were consumed in 1910. Both Mexico and Australia each registered around three hundred

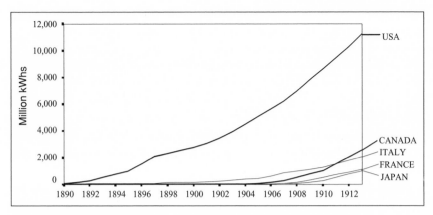

FIGURE 3.2. Five Largest Producers of Hydroelectric Power, 1890–1913
Sources: See Appendix A.

million kWhs consumed per year before the First World War, and even China consumed an estimated fifty million kWhs in 1912.[43]

In most countries, the bulk of this electricity was generated by burning coal in steam engines and sophisticated turbines. For this reason, the spread of electrical systems spurred even greater demand for coal. However, the development of turbines also allowed water power to be harnessed in new ways. As a result, the emergence of modern electrical grids was associated with the large-scale incorporation of another primary form of energy— that of running water—into the modern global energy system.

Given American advances in these electrical systems, it is no surprise that the first large-scale hydroelectric plant was opened at Niagara Falls, New York, in 1881. The successes of the Niagara project spurred new projects throughout the United States, so that by 1913 hydroelectricity was providing more than 40 percent of the country's electricity. As Figure 3.2 shows, the United States outstripped all other countries in hydroelectric energy production before the First World War. Neverthe-less, steady expansions were achieved in hydroelectric sectors in Canada, Italy, France, and Japan in this era as well.[44]

Clearly, the decades leading up to the First World War saw important advances in the field of electrical power in general, and hydroelectricity in particular. Just as with the earlier growth of oil-based energy systems, this initial phase of expansion of electrical systems was spurred by a con-vergence of commercial and social factors, rather than being driven by material demands. And the speed with which complex, expensive electrical grids spread across the industrial world reveals yet again that the modern global energy system has an amazing capacity to absorb new technologies and resources in a rapid and far-reaching manner.

Declining British Hegemony and the Expansion of New Energy Systems

The early twentieth century witnessed the steady expansion of oil and electrical systems throughout the advanced industrial world. Given that these new energy systems were overlaid on top of a still-growing coal regime, businesses and citizens of core countries came to enjoy historically unprecedented levels of energy consumption during this period. With increasing access to coal heat, oil-powered transportation systems, and electric lights, urban citizens across North America, Western Europe, and some other areas began to experience profound transformations in their lives. Manufacturing plants could operate at close to full capacity at all hours of the day, and during the entire year. Retail stores could stock their shelves and display their wares more effectively. Urban residents could make use of an array of new transportation options, home heating systems, and lights to broaden the material and cultural horizons of their lives.

During the first decade of the twentieth century, therefore, the global energy system fueled a massive expansion in capitalist production and consumption. However, beneath these positive developments, ominous trends were also emerging. For instance, death rates in underground coal mines climbed steadily. In the United States alone, over one thousand miners died each year in coal mines during the century's first decade. In fact, the year 1907 alone witnessed the death of over three thousand coal miners in U.S. mines.[45] Mortality also rose on oil rigs, utility lines, and increasingly congested city streets. Meanwhile, the air in many cities in Europe and North America grew dense with smog from coal and oil combustion. Clearly, energy systems began exacting significant human and environmental tolls even before World War I.

Most ominous, however, was the increased lethality that an emergent oil system offered to military forces across the world. Although it was not immediately apparent to most citizens, a shift of historic significance was under way. By the first decade of the twentieth century, military strategists in many core nations were experimenting with new combat uses for oil. Following successful experiments, advanced industrial states in Europe and North America began to intervene directly in oil-related sectors. Even before the onset of the First World War, yet another modern energy resource was being incorporated into the world's most advanced weapons programs.

The use of oil-based systems in military campaigns first occurred in colonial conflicts. As has been discussed, coal-powered steam boats and railroads played a key role in allowing European powers to conquer new areas in Africa and Asia during the late nineteenth century. By the first decade of the twentieth century, oil-powered vehicles and diesel-powered boats were also being used in these colonial campaigns. For example, airplanes were used in colonial wars in Algeria, Morocco, and Turkey in the period 1911–13.[46] As with coal, oil intensified the strategic imbalance that existed between core and noncore countries in the twentieth century.

Of much more serious consequence, though, was the way in which oil would be used to increase the destructiveness of wars between the world's most powerful nations. In the first decade of the twentieth century, Britain began to lose its dominance as the world's hegemonic power. As we have seen, the United States and Germany emerged as pioneers in new petroleum and electrical industries. These nations also undertook their own military campaigns, with the United States seizing new territories in the Caribbean and the Pacific and Germany moving into Africa and the Middle East. Rising military tensions also emerged in the core of the world-economy itself, as Germany adopted an increasingly aggressive military posture against France and England.

The intensification of geopolitical competition associated with Britain's hegemonic decline spurred state intervention in emergent oil sectors. This intervention was first manifested in naval operations, and then it spread to aviation and automotive sectors in the prewar period. Great Britain pioneered the use of oil in ships during first decade of the 1900s. This is intriguing, since Britain had almost no domestic oil supplies at the time. The British Admiralty was divided over the wisdom of embarking on a full-scale shift to reliance on oil. Many British strategists argued that national security would be undermined by a shift away from coal-fueled ships, since coal was so plentiful in the British Isles. Nevertheless, there was growing evidence that oil-powered battleships offered superior fighting capabilities. Oil-powered vessels were easier to refuel, achieved cruising speeds much more quickly than coal-powered ships, and produced less visible exhaust on the open sea.

The British Navy initially decided to deploy small ships that incorporated oil-fueled turbines. After successes with these trials, the decision was made to build oil turbines into the nation's most ambitious naval ship, the Dreadnought. Completed in 1906, the Dreadnought immediately rendered all other battleships obsolete. The successful use of oil in this ship gave added confidence to those who advocated a full-fledged shift to reliance on petroleum. And, indeed, by 1912 it was decided that all new U.K. battleships would burn oil.

British advances in oil-based naval systems spurred similar efforts in other core nations as well. The navies of Germany, Italy, Russia, and the United States also began to use oil in new ships in the first decade of the century.[47] Indeed, by the onset of the First World War, virtually all leading navies were in the process of converting to oil. In addition, most of these countries began purchasing oil-powered vehicles and aircraft as well.

Growing military reliance on oil prompted European governments to increase their investments in foreign petroleum sectors before the war. The result was the creation of a new group of hybrid oil companies, which were part private and part public entities. The decision by the British Navy to convert to oil, for instance, provided a strong stimulus to efforts to develop multiple sources of oil. Within a few years of its decision, the British Admiralty was contracting with companies to develop oil sources in India, Burma, Mexico, Romania, and the Middle East.

The first of these joint ventures began in 1905, when the British government signed a long-term contract with a small company, Burmah Oil, for the provision of naval oil supplies. In support of their new partner in oil, the British authorities prohibited all foreign companies from producing or selling oil in India, thereby excluding Standard Oil from one of its largest export markets (Jones 1981). With a guaranteed large customer, protection from competition, and regular infusions of public capital, Burmah Oil expanded across Burma and India. Dutch capital, for its part, focused on developing oil wells in the East Indies. Meanwhile, companies from France and Germany struggled with less success to gain access to foreign reserves.

European efforts to gain control over foreign oil supplies drew them inexorably to the Middle East, where mammoth discoveries were being made. The British government was at the forefront of this effort to develop secure access to Middle Eastern oil reserves. In 1910 Burmah Oil and the British government invested in a second venture, this time in Persia. This new venture, soon named the Anglo Persian Oil Company, grew out of a concession that had been negotiated by William D'Arcy, a British merchant, and the Grand Vizier of Persia in 1901. Having been granted the exclusive right to explore for oil in 480,000 miles of Persian territory, D'Arcy soon ran into difficulties and was forced to turn to Burmah Oil and the British government for financial support and market outlets (Ferrier 1982). Although nominally a private company, in reality the British government came to own the majority of its shares, the British Navy became its main customer, and Anglo Persia received diplomatic and military protection for the next half century.

The Anglo Persian Company remained an exclusive British concern. However, the other crucial venture undertaken prior to World War I in the Middle East—the Turkish Petroleum Company—received investments from British, Dutch, and German companies. German companies also invested in new drilling operations near Mosul, in an area that would become one of the world's richest sources of crude oil. Additional German investments went into rail and port facilities that had, as part of their intended purpose, the eventual transport of oil to domestic consumers. French capital was also invested in foreign oil ventures through participation in Royal Dutch Shell projects in the East Indies, Russia, and elsewhere.[48]

Overall, the conversion to oil by the world's leading navies provided an important early stimulus to demand for this new fuel. At the turn of the century oil was generally between three to twelve times as expensive as coal in most European and North American markets, which limited its appeal to private consumers. Because of large-scale, prewar naval conversion programs, however, demand for oil stabilized. Technological hurdles to burning oil in large and small engines were also ironed out by military engineers.

Heightened geopolitical competition associated with a decline in British hegemony clearly caused governments across Europe, North America, and beyond to increase support for oil and electrical industries in the period 1880–1913. This state support helped private companies scale up their operations, so that embryonic energy systems based on oil and hydroelectricity grew steadily in the early part of the century. At the same time, rising labor unrest in coal industries caused periodic disruptions in the production and marketing of this resource. Under the combined impact of these political, commercial, and social pressures, the world was already witnessing the consolidation of two new energy systems in the pre–First World War period. As we shall see, the chaotic impact of two world wars had the unexpected effect of accelerating the expansion of the global energy system based on petroleum in the following decades.

4 The First Period of Crisis

I n the first decade of the twentieth century, the world appeared to be entering a period of prosperity fueled in part by an unprecedented abundance of energy resources. Even though conflicts between coal mine owners and workers often swept through the world's main mining regions, coal production continued to spread steadily into new regions of the world. At the same time, the incorporation of the new resources of oil and hydroelectricity significantly increased the power available to citizens in the developed regions of the world. Up through 1913, in short, the capitalist world-economy appeared capable of generating almost continual expansions in energy production and consumption.

However, modern energy systems bring with them more than progress and material advancement. These systems also radically intensify the destructive capabilities of military forces. These facts became clear as the world-system shifted from a period of relatively peaceful expansion under British hegemony into an era of severe conflict.

This tumultuous period witnessed a profound shift in world energy industries. Many centers of coal production were damaged in military campaigns, while governments shifted their support strongly toward oil. Coal companies were also confronted with new waves of labor unrest, which complicated their efforts to compete commercially with rising oil corporations. As a result of this convergence of political, commercial, and social factors, oil companies were able to establish commercial dominance over the older coal system in the period 1913–46. An investigation of this period shows again that, if global conditions are right, fundamental transitions can occur in the foundations of the global energy system in a matter of decades.

The Impact of World War I on Global Energy Industries

Although political tensions between European states grew steadily during the first decade of the twentieth century, there was little forewarning of what kind of conflict was brewing. It was widely expected that, if war came, it would follow the pattern of recent conflicts in Europe. The Crimean War, the Franco-Prussian War, and the Russian-Japanese War had all been short conflicts that had only temporary impacts on resource stockpiles. European armies therefore did not amass large quantities of coal or oil as they prepared for conflict. Instead, military planners assumed that they could either fuel their forces from secure sources or else capture the energy resources that were needed to prosecute their war campaigns.

When World War I began, coal fields quickly became targets of military campaigns. One of the central objectives of French military commanders was to recapture the coal-rich regions of Alsace and Lorraine that they had lost in the Franco-Prussian war. German strategists, meanwhile, were determined to seize key coal fields in Belgium and Poland. As it turned out, Germany was at first most successful in carrying out its objectives. Not only were German forces able to seize mines in Belgium and Poland early in the war, but they even penetrated into France and captured the mines of the Nord.

Once the western front had slowed into a stalemate, one set of trenches ran through the key French coal region of Pas-de-Calais. The result was that approximately two thirds of France's coal mining capacity was destroyed. Later, as German forces were repelled from captured territory, coal mines in Poland and Belgium suffered heavy destruction as well.[1] In addition to the damage wrought in these important coal regions, there was a heavy toll on an entire generation of miners. Workers from Germany, France, and Britain were sent to the trenches, where it was hoped that their tunneling expertise could help break through the stalemated western front. Over the long years of trench warfare, tens of thousands of European miners were killed or severely wounded in these subterranean campaigns.[2]

The war clearly had a devastating impact on key coal mining industries in Europe. At the same time, the war stimulated a rapid expansion in oil-powered industries. At first, oil-burning vehicles played only a limited role in military campaigns. As the war dragged on, though, military strategists turned to tanks and airplanes in efforts to break through the stalemate on the western front. Meanwhile, diesel-powered destroyers and submarines engaged in new forms of naval combat on the high seas. The result was a massive increase in the production of oil-burning vehicles, ships, and airplanes. For instance, the world tonnage of oil-powered vessels more than tripled between 1914 and 1920. Meanwhile, global production of gasoline-powered vehicles grew by about a factor of five over the same period, and airplane production in Europe and North America grew exponentially during the war.[3]

As the war progressed and oil-powered vehicles became increasingly important in combat, efforts to gain access to oil supplies grew more intense. Germany fought with Britain for control over the Baku region, for instance, while Britain sent troops to protect its oil operations in Persia. Meanwhile, warring navies tried to stop shipments of oil to their opponents. As oil deliveries from the United States increased, Germany unleashed a wave of submarine attacks on shipping lines. Germany was not able to stop oil imports to Britain and France, though, and instead the attacks prompted the United States to enter the war. American oil exports to Britain and France then soared, while German access to oil was almost completely severed. The fact that U.S. multinational companies had supplied almost all of Germany's oil before the war made it easier for this embargo to be implemented.[4]

Wartime industrialization efforts also stimulated the expansion of electrical systems. Factories in the United States and Germany, in particular, began making greater use of hydroelectricity as coal resources became scarcer due to labor difficulties. Hydroelectric production also underwent growth in France and Italy as imports of oil were disrupted.[5] State-sponsored programs to expand long-distance transmission lines from sites with water power to industrial centers intensified during the height of the war effort. As a result, electrical utilities emerged from the war with a strengthened position in manufacturing sectors in much of the developed world.

While systems based on oil and electricity were stimulated by the First World War, coal emerged from the war as an industry in crisis. A whole generation of European miners was killed or injured, and mines all across Europe were severely damaged. In addition to these direct impacts, the pressures of war generated labor/management tensions that exploded toward the end of the war. In fact, a wave of labor militancy rippled through virtually all major coal sectors in Western Europe and North America in the aftermath of the First World War, which helped erode confidence in the industry and accelerate a shift toward oil in subsequent decades.

The country that witnessed perhaps the most disruptive wave of labor unrest was Britain. During the war, miners who were allowed to remain in domestic mines rather than fight were forced to increase their work effort. With this intensification, injury and death rates within British mines also escalated. As a result, industrial unrest grew so intense that the government was forced to intervene and nationalize virtually all coal mines in 1917. During the wartime period of government control, strikes were outlawed. In the aftermath of the war, however, from 1919 to 1921, massive strike waves swept through the industry.[6]

The First World War also generated an intense wave of labor/management conflict in France. Before the war, French coal miners had forged a strong union movement that had won significant improvements in wages and working conditions. During the war, however, virtually the entire membership of the mine union movement in France's northeast coal region was killed and the union organizations collapsed. In the context of wartime pressures, French mine owners and state authorities turned to increasingly authoritarian efforts to maximize coal output in the mines that remained in operation. Though perhaps understandable in the context of the war, these management practices persisted after the conflict was over. In response, miners in regions that had been spared the destruction of the war spearheaded massive strikes in the years 1919–20.[7] The French coal sector therefore emerged from the First World War as a crisis-ridden industry.

Much the same pattern occurred in Germany and Belgium as well. A massive strike wave engulfed the Ruhr and other coal regions in 1919–20, resulting in widespread violence and military intervention.[8] In Belgium, a similar wave of unrest swept through coal mines from 1919 to 1921. In both countries, national confidence in domestic coal sectors was undermined and the countries shifted more resolutely toward oil and hydroelectric resources.

Even in North America, which was spared the worst of the war's destructive impacts, major waves of labor unrest spread through coal sectors. In the United States, for instance, a major strike took place in 1919 that shut down around 75 percent of the industry. An even larger battle between workers and managers was fought in 1922, in what proved to be one of the most disruptive coal strikes in U.S. history. The strike was successful in shutting down production in wide segments of the U.S. coal industry for many months and in generating coal shortages and sharp increases in prices across the country. In Canada, massive coal strikes also broke out in the period 1917–20. After causing significant economic turmoil, the workers won large wage concessions.[9] In each country, these conflicts between labor and management provided an added impetus for consumers to shift toward oil for their energy requirements.

Intensifying Systemic Dynamics and the Expansion of Oil

With their coal industries in crisis and demand for oil rising in commercial and military sectors, the world's leading governments emerged from the war with an intense determination to gain access to oil reserves. Because they lacked domestic oil supplies, European governments and companies in particular began a concerted effort to seize control of foreign oil supplies as soon as they were able to do so.

A major new opening occurred with the collapse of the Ottoman Empire at the end of the First World War. Because the empire had allied itself with Germany and the other Axis powers during the war, it was treated as a defeated enemy in postwar negotiations. The European victors, driven by a new determination to gain access to the region's oil, fought with each other over the dismemberment of the Ottoman Empire. Within a few years British and French companies had seized control over most of the known oil fields in the region, and they managed these resources with little consultation from indigenous authorities.

The period following the First World War marks the beginning of a period of imperial penetration that fundamentally reshaped Middle Eastern societies and economies. The countries in the region often came to define their borders only when the oil companies required maps delineating their concessions, leading to mismatches between national and cultural boundaries that continue to create difficulties today (Hourani 1991). Furthermore, the incorporation of dynastic ruling families into oil-based international financial networks exacerbated income inequalities and insulated ruling elites from popular accountability over the long term, producing the authoritarian regimes prevalent in the region today. Finally, the strategic importance of the region's oil resources drew leading powers to intervene repeatedly in Middle Eastern political affairs, setting the stage for acrimonious clashes in later decades.

One of the main reasons for British intervention in Persia (present-day Iran) was the fact that the Admiralty had become a major stockholder in the Anglo Persian Oil Company. Anglo Persia had run into difficulties in refining and marketing the oil that was coming out of Persian wells. The heavy oil that was extracted was unsuitable for civilian markets, but it could be used to power ships. Anglo Persia found its commercial salvation in the British Navy, which provided capital, a huge guaranteed market, and political protection from regional unrest. In exchange, the Admiralty was able to obtain oil supplies from the Anglo Persian Oil Company at far below market prices. Indeed, Anglo Persia became the navy's single largest supplier, providing over 50 percent of the navy's oil during the next two decades (Jones 1981). Britain's naval strength therefore became intimately linked to the exploitation of the Persian oil fields.

Although Britain established an early dominance in the region, government and corporate interests from other countries also struggled to gain access to Middle Eastern oil reserves during the next sixty years.

A complex series of commercial agreements were designed and then re-peatedly revised, as broader geopolitical relationships between the world's most powerful nations underwent major shifts. As a result, from the end of the First World War through the early 1970s, dominance by Britain gave way first to joint control between Britain and France and then to increasing participation by U.S. companies.

The commercial negotiations initially reflected the influence that Britain and France had in the Middle East at the end of the First World War. The San Remo Agreement of 1920, for instance, granted British and French companies privileged rights to exploit the rich oil fields of the Mesopotamian region. At the same time, Dutch interests controlled most production in the East Indies while Britain controlled Indian and Burmese oil sectors.[10] As a result, in the early 1920s U.S. companies found themselves formally excluded from many of the world's most prolific oil regions.

Successive discoveries of large oil reserves from Pennsylvania to Texas to California had initially lulled U.S. officials and companies into compla-cency regarding international oil issues. However, the announcement by the U.S. Geological Survey in 1919 that domestic oil reserves might be depleted within a few decades greatly alarmed naval authorities and oil en-trepreneurs. This new perception of scarce domestic oil reserves prompted the U.S. federal government to enter into an unprecedented joint effort with private oil companies to push for access to foreign oil reserves. By the early 1920s the federal government was exerting its diplomatic influence in the Middle East on behalf of American oil companies.[11]

U.S. efforts to penetrate the Middle East were initially rebuffed by Britain and France, who controlled virtually all known reserves. However, the British and French soon found that they were overextended in the region. By the mid-1920s armed revolt in Iraq was challenging British rule, while an emerging Turkish-Russian alliance was threatening western interests in Iran (Stivers 1982). In an effort to dominate the region, Britain unleashed air strikes against rebellious groups in Iraq and increased the revenues flowing to pro-Western local elites in Egypt and Iran.

In response to this British offensive, leaders in Turkey, Persia, Iraq, and Egypt began urging U.S. President Wilson to apply his principles of self-determination to the Middle East. In exchange for formal independence and diplomatic recognition, Arabian rulers announced that they were will-ing to break the concessions they had signed with the Europeans and sign favorable oil deals directly with U.S. companies. However, the Wilson administration refused to negotiate with local elites and instead pushed for admittance to the European-designed concession agreements as a

full-blown imperial participant. In the process, the United States alien-
ated an entire generation of Middle Eastern leaders.[12]

Faced with armed unrest in the Middle East and a diplomatic offensive
from the U.S. government, the Europeans eventually allowed American
companies to negotiate entrance into the region's oil concessions. The
Red Line agreement, signed in 1928, assigned British, French, Dutch, and
American oil companies specific drilling rights in Turkey, Persia, Iraq, and
later Abu Dhabi, Qatar, and Saudi Arabia. A complementary marketing
agreement, signed at Achnacarry Castle in 1928, fixed international oil
prices, sought to control excessive competition by assigning each oil major
a portion of key markets, and facilitated coordinated offensives against
independent companies trying to break into the Middle East.[13]

The conclusion of the Red Line and Achnacarry agreements ended a
period of intense competition among the major international oil companies and their home governments. The compromises provided a stable
political environment within which oil capital from Britain, the United
States, France, and the Netherlands could operate with confidence. As the
1920s drew to a close, oil production began to expand with great rapidity
in Iran and Iraq. Major new oil discoveries in the western United States,
Venezuela, and Romania in the late 1920s drove global oil production capacity radically higher. As these resources began to be shipped to major
markets in Western Europe, North America, and other parts of the world,
oil companies intensified their efforts to win market share away from established coal industries.

Commercial Competition between Coal
and Oil during the Interwar Period

On the eve of the First World War, oil companies began to diversify out
of their original markets and into a variety of new sectors. By the end of
the Second World War, it was clear that oil companies had triumphed
over their coal competitors not only in emerging automobile and aircraft
industries, but also in long-established coal markets in rail and shipping
sectors.

It is important to emphasize that this penetration of oil into new markets
was not an automatic or inevitable process, but was instead carried out in
the face of many political, commercial, and social hurdles.

One key difficulty facing oil companies in the interwar period was the
fact that oil prices were consistently higher than those of coal. Prices for
specific energy commodities vary substantially from city to city and from
country to country, so comparative price data can only present the broadest

of relationships. Nevertheless, it is clear that a substantial price differential between coal and oil persisted into the post–First World War era in most places.

In the United States, which was the only market-economy with large coal and oil industries, oil prices reached the closest parity with those of coal. As shown in Figure 4.1, in 1880 the average U.S. price of crude oil had basically matched the price of anthracite coal, and by the early 1900s had become about 30 percent cheaper than anthracite. However, the average price of bituminous coal remained substantially below that of oil into the 1950s. Since bituminous coal supplied over 60 percent of the energy requirements of the United States in the interwar period, this low-cost fuel was a major competitor to oil. Even in the United States, where resource endowments most favored oil companies, these firms clearly had to compete against low-cost coal resources in most markets.

Turning to Western Europe we see that the price differentials were even more substantial, largely because imported oil faced long-established domestic coal industries. In both Britain and France, for instance, oil prices were at least four times as expensive as coal into the 1920s. Even as Latin American and Middle Eastern imports began to flow into Britain and France after the First World War, its price still remained higher than that of coal. For the rest of Western Europe, oil prices were three to four times higher than coal throughout the inter war period, only reaching parity in the mid-1950s.[14] In Western Europe, as in the United States, oil firms had to contend with significant price differentials during the interwar period. The fact that oil companies were able to triumph commercially in the face of these price differentials reveals that energy shifts are not determined by simple cost differences.

There was another difficulty that emerged at the turn of the century as well: the unreliability that plagues all new technologies. As with steam engines a century earlier, frequent explosions in an early generation of oil-powered engines limited the extent to which private firms were willing to shift to the new fuel. Coal companies often seized upon these oil-fueled accidents, urging consumers to stick with long-established steam-based systems. Coal advocates in France and Belgium were particularly effective in slowing the shift by private industries to oil in the interwar period, while similar efforts were mounted in Britain, Germany, and the United States as well.[15] Nevertheless, oil companies were able to overcome these multiple commercial hurdles and achieve remarkable growth during the interwar period. This again reveals that the natural tendency for consumers to mistrust new energy systems is not an insurmountable barrier to the adoption of novel technologies and resources.

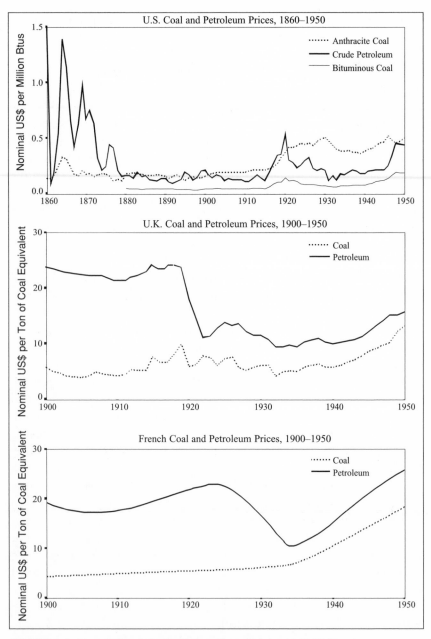

FIGURE 4.1. International Coal and Petroleum Prices, 1860–1950
Sources: U.S. Department of Commerce (1975), Ferrier (1982), Mitchell (1988),
Bairoch (1993).

Part of the explanation of the success of oil companies lies in the heightened levels of state support that oil-based systems received during the First World War and afterwards. As discussed earlier, wartime production drives were key to improving the reliability of oil propulsion systems and scaling up the manufacturing of new vehicles. While state support for oil-based technologies diminished after the war, it certainly did not disappear. Throughout North America and Western Europe, for instance, public-private ventures worked to expand the civilian road networks required by the new automobile. In the United States these efforts began in 1916 with the passage of new gasoline taxes earmarked for highway construction, and then expanded with the Highway Act of 1921 and the public works projects of the New Deal era. Similar projects emerged in Western Europe during the interwar period as well. Adolph Hitler, in particular, spearheaded a major highway construction program in Germany in the 1930s. While denser urban centers allowed mass-transit systems to grow more easily in most European countries, intercity traffic increasingly shifted toward autos and trucks. International airline routes flown by U.S. and European companies also received state subsidies during the interwar period.[16] Overall, state policy began to shift increasingly in favor of oil-based transport systems in many advanced economies in the mid-twentieth century.

Although governmental support was important, it alone cannot fully explain the successes achieved by oil companies in many different market environments. For a more complete explanation, which adequately accounts for the commercial dimension of the transition, it is necessary to turn to a comparative analysis of oil and coal business enterprises. Although there were numerous kinds of firms in oil and coal industries throughout the world and competitive struggles therefore varied significantly, it is still possible to highlight the central factors that pushed oil companies to evolve in directions that enhanced their advantages vis-a-vis coal firms.

As has already been discussed, the major oil companies spent decades developing networks for the production and distribution of illuminants and lubricants before they entered into direct competition with coal firms. In this early period of evolution, the oil majors were faced with three unique challenges that were not present in coal. Most oil firms failed to negotiate these hurdles successfully, but the few that did were forced to develop organizational strengths that then helped them out-compete coal firms.

The first challenge faced by oil companies was the need to develop refining processes that would allow many different kinds of crude oil to be transformed into standardized commodities. Whereas coal companies generally sell their fuel in its original form, oil companies had to develop increasingly sophisticated refining systems to deliver standardized, reliable fuels to consumers. In contrast with coal companies, therefore, the major oil companies were pushed to develop extensive research and development divisions.[17] This technical requirement was one factor favoring the emergence of new, vertically integrated businesses in oil industries.

These refining activities led to a second challenge. In the course of processing crude oils, numerous byproducts are generated that were initially treated as waste. To improve profit margins in their refineries, however, oil companies were compelled to find markets for this growing variety of oil derivatives. At the turn of the century Standard Oil, for instance, began promoting such products as gasoline, naphtha, paraffin, solvents, and oil jelly in addition to traditional fuels like kerosene. This early focus on marketing provided important learning opportunities for oil firms, helping prepare them for the period of direct competition with coal.

Efforts to promote new oil products in coal's primary markets date from the turn of the century, following the massive oil discoveries that took place in the United States and Latin America. After 1901, for instance, Gulf and Texaco began offering discounts to shipping and railroad companies in the western United States, where oil production was booming and coal was less prevalent. Similarly, campaigns were carried out by Royal Dutch Shell to convince shipping companies in Europe to switch to fuel oil. Breakthroughs in gasoline refining processes achieved by Standard Oil scientists in the 1910s, meanwhile, brought this group of companies into gas and fuel oil markets as well.[18] By the second decade of the twentieth century, gasoline and fuel oil had replaced kerosene as the main products coming out of refineries, signaling the onset of large-scale competition between coal and oil. By the time this direct competition began, the major oil firms were able to draw on their extensive marketing experience to devise strategies of expansion.

The third challenge facing oil companies, which distinguished them yet again from their coal counterparts, was the need to continually expand the geographical scope of their operations. Faced with the continual possibility that reservoirs could run dry, oil companies were compelled from an early period to search for new reserves. The major oil companies attempted to minimize the risks that rival firms would gain access to low-cost oil, or that political and social unrest would disrupt supplies, by continuously diversifying their sources.[19] Therefore, even after the major oil companies had developed a surplus of reserves, they continued to enter into ventures in new areas.

As a result of these expansionary pressures, the organizational characteristics of oil came to differ significantly from those of coal. Coal

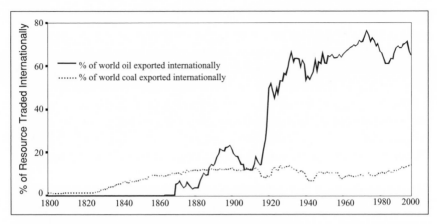

FIGURE 4.2. Export Ratios for Coal and Oil, 1800–2000
Sources: See Appendix A.

firms typically remained tied to particular ore bodies. Even when mining capital was invested across international borders, this was usually carried out through the creation of largely independent firms. By contrast, major oil companies were forced to shift their operations from region to region, and so they developed into geographically diversified businesses. Indeed, the oil industry became internationally oriented more rapidly and deeply than did coal. As shown in Figure 4.2, the proportion of coal that was traded across borders grew slowly, stabilizing at around 15 percent. Meanwhile, oil's trade ratio surpassed that of coal a mere three decades after the industry's inception, and eventually came to average somewhere between 60 and 80 percent. Much of the international oil trade was carried out through transfers between subsidiaries of the major oil companies, and so this international trade reflects to a large extent the internationalization of the major oil corporations.

The international orientation of the major oil companies had a tremendously important role to play in their eventual triumph over their coal rivals. In particular, the major oil companies could rely on extremely large profits generated in their foreign operations to subsidize their activities in more competitive European and North American markets.

Although data on oil company profit rates have to be treated with caution,[20] the information we have nevertheless reveals that foreign oil operations generated remarkable returns. During the 1920s and 1930s, the domestic operations of U.S. oil companies returned annually somewhere between 4 and 15 percent profit on total investments. In comparison, Standard Oil of New Jersey's Venezuelan subsidiary, Creole

Oil, generated such massive returns that this single subsidiary regularly produced over 40 percent of Jersey Oil's total annual profits in the 1920s. Profit rates generated by Anglo Persia, meanwhile, averaged around 20 percent from 1918 through 1938. Finally, the annual dividends paid by Burmah Oil and Royal Dutch Shell exceeded 30 percent during the same decades.[21]

By contrast, profit rates in major coal industries were extremely slim. In Britain, for instance, wage costs jumped dramatically after 1914, reducing average coal company profits to single-digit levels. In both France and Germany, most coal companies survived the destruction of the First World War and the Great Depression only as a result of extensive government assistance. From 1929 through 1939, meanwhile, U.S. bituminous companies registered net losses nearly every year.[22] While most coal firms were struggling to survive, internationally integrated oil companies were able to transfer resources from their lucrative foreign operations to sustain their activities in more hotly contested markets in North America and Europe.

The three competitive challenges faced by oil companies—the need to develop refining systems, the need to aggressively market a variety of products, and the need to continuously expand their terrain of operations—favored the emergence of a small number of vertically integrated, multinational oil companies in the interwar period. In the United States, for instance, some fifteen companies had come to control over 60 percent of the country's oil acreage by 1926. On an international level, the dominance of such firms was even more apparent. By the 1920s seven oil companies—Standard of New Jersey, Standard of New York, Standard of California, Royal Dutch Shell, Texaco, Anglo Persia, and CFP—had come to dominate the global trade in oil.[23]

It is important to point out that the majority of oil companies were unable to overcome the various challenges posed by their particular business, and most therefore remained confined to particular regions or lines of production. Even for those oil companies that did manage to evolve into powerful, multidivisional enterprises, the process of organizational change was quite difficult.[24]

Similar to their coal counterparts, the major oil companies generally began their existence as family-run, centralized businesses. Standard Oil, for instance, was controlled by Rockefeller, his relatives, and then a small number of directors into the 1920s. Shell Oil originated as a firm run by the Samuels family, and even after the business merged with the Royal Dutch Company, relatives of the family continued to manage Shell operations. Royal Dutch, meanwhile, was dominated by Deterding until the late 1930s, while Anglo Persia was controlled by a select group of Scottish investors well into the 1950s.[25] In each of these companies, early entrepreneurial successes encouraged an aging generation of directors to attempt to run virtually all aspects of their businesses.

As the firms diversified into new markets and activities, however, these centralized lines of authority became overburdened. Even after it had become clear that organizational changes were needed, major reforms came only after lengthy internal struggles within each of the major companies. Though given an initial push as a result of the 1911 dissolution, for instance, Standard Oil's offshoots struggled for the next two decades to institute decentralized, multidivisional forms of organization. Similar corporate reorganizations were carried out within Gulf Oil, Texaco, Phillips, and a host of other U.S. oil companies during the 1930s. Royal Dutch Shell and Anglo Persia, meanwhile, had to go through periods of serious crisis before a new generation of directors, attuned to the need for more flexible management structures, gained control.

For their part, coal companies had little obvious reason to carry out such difficult organizational changes. With access to large deposits of coal and accustomed to selling a single product to a relatively stable group of customers, these firms were not generally forced to carry out aggressive policies of geographic expansion, product development, or sales promotion campaigns. Even when British, U.S., Australian, or South African coal companies increased their international sales after the First World War, these transactions were generally carried out on the basis of long-term contracts that did not require the creation of foreign subsidiaries.[26] As a result, coal companies were not faced with particularly intense pressures to undergo the difficult transition from centralized, often family-run firms to multidivisional, transnational businesses.

It is also worth noting that crossover investments between coal and oil sectors slowed efforts by coal interests to protect their market position in numerous countries. For instance, the Rockefeller company invested heavily in coal companies and railroads, so that by the first decade of the twentieth century this oil corporation controlled large portions of the coal industry in the Western United States (Seltzer 1985: 37). This measure of control forestalled serious defensive campaigns by coal companies in western states, where oil rapidly captured markets.

An extensive literature has castigated European and North American coal owners for failing to adequately respond to the competitive onslaught of oil companies during the interwar period.[27] There certainly were worrisome trends emerging in most coal sectors. Productivity gains in coal fell behind those achieved in oil, and dynamics of ruinous competition persisted in many countries.[28] But it is not at all clear that these problems stemmed from entrepreneurial failure. Instead, the combination of many factors—rising labor militancy in coal, the inability of coal companies to control competition, the amassing of financial strength by oil companies in peripheral countries, and the development of new organizational strengths by the major oil companies—facilitated the shift toward oil during the interwar period.

Conflicts over the Organization of Work in Coal

While oil companies mounted aggressive campaigns to win market share away from coal, the coal companies themselves became embroiled in a complex effort to deal with the labor militancy that had become endemic to the mining industry. As has been demonstrated, waves of labor militancy emerged in the world's key coal mining sectors in the 1880s, and then intensified in the period 1919–21. Early in the interwar period, then, coal companies throughout the world were faced with the challenge of containing this unrest and reestablishing profitability in the industry.

There was a variety of possible responses to this rising labor militancy in coal. The immediate response from coal owners and state agents tended to be overt repression, but the use of violence frequently led to even larger protests, and as a result alternative strategies began to be developed in response to the labor unrest.

The first strategy that was utilized in a systematic fashion during the interwar period was the shift toward mechanized underground coal mining. This process was partly spurred forward by the First World War. Faced with an escalating demand for the crucial war resource and chronic labor shortages, mine operators in Europe and North America turned increasingly to the use of mechanical cutters and conveyors in their underground shafts. This shift in work procedures not only held out the possibility of increasing productivity, but it also provided an opportunity to reduce the autonomy of the miners laboring underground.

The use of mechanized rock cutters required that miners shift from traditional room-and-pillar methods of mining to what became known as long-wall mining techniques. In the original room-and-pillar approach, small groups of skilled miners cut tunnels through coal seams, leaving pillars and walls in place to hold up the roof. As a result, the skilled miners spent most of their time working in their own isolated chambers. Supervisors were hard-pressed to visit all the chambers, and so the workers had a high level of autonomy at the coal face. The miners were also generally in charge of making their own decisions as to where to cut and where to leave protective walls in place. The miners therefore had a certain degree of control over the level of danger they were exposed to from cave-ins.

The introduction of mechanized cutters and conveyor belts changed many of these characteristics of underground mining. The machine cutters, for instance, required the excavation of long, relatively straight passages through the rock. Moreover, the mine chambers had to be expanded to accommodate the equipment, and the workers had to be gathered together into these larger chambers, so that the flow of ore from the cutters to conveyor belts and up to the surface could be more effectively coordinated. As miners were moved out of their individual chambers and into centralized work areas, they were exposed to higher levels of surveillance. Supervisors therefore strengthened their ability to ratchet up the intensity of work in the underground shafts.[29]

In addition to being subjected to more intense levels of surveillance and work, the underground miners were also exposed to new dangers with the introduction of mechanized cutters. When the miners had relied on pick axes to cut the coal, they were able to listen for creaks in the roofs above them to get some indication of when a cave-in was likely to occur. This strategy could no longer be used once noisy machinery was introduced into the shafts. Furthermore, the machines generated a huge volume of harmful dust as they ate into rock walls and veins of ore. Finally, the miners often lost control over which rock formations would be cut and which would be left behind as support pillars. Rather than making their own decisions within the shafts, miners were generally directed to mechanically cut paths along lines drawn by managers who worked above ground. As a consequence, many miners came to feel like life–and–death decisions were being made by people who had little understanding of particular rock formations—and who would not themselves die if maps turned out to have left inadequate support systems in place.[30]

Given that mechanization generally reduced the autonomy of workers and increased the risks associated with underground mining, it is no surprise that miners were less than enthusiastic about the introduction of the new techniques. Indeed, protracted struggles took place during the interwar period between miners and owners in Western Europe and North America over processes of mechanization. Miners did have some success in delaying the speed of mechanization, which limited the ability of coal companies to cut prices in their competition with oil. However, they were not able to stop that advance of mechanization altogether.

Coal miners in Britain, for instance, emerged from the First World War determined to protect their traditional work procedures, but they were confronted by company owners and government officials who were equally determined to increase the productivity of the industry. After a decade of struggle, in which they rather successfully resisted major changes in work procedures, the miners suffered a defeat during the strike of 1926. From that point forward, mechanization proceeded at a rapid pace in the United Kingdom. Similar patterns occurred in France, Germany, and the United States. In each case, defensive strikes were carried out during the 1920s, but once the Depression hit, miners were forced to accept work under almost any circumstances, and so mechanization accelerated in these countries.

By the beginning of the Second World War, important shifts toward mechanization had taken place in all these countries. While at the beginning of the First World War only about 8 percent of coal was mechanically cut in Britain, by the late 1930s that proportion had risen to approximately 60 percent. In France and Germany, virtually all coal was cut by hand up through the First World War, but by the beginning of the Second World War over 50 percent was cut with machines in each country. In the United States, the percentage of underground coal cut by machines rose from under 40 percent to almost 90 percent between 1913 and 1935.[31]

Another strategy for countering labor militancy—that of moving to regions where unions were nonexistent—was also used in the United States during the interwar period. For instance, studies have shown that coal company investments shifted away from unionized mines in Pennsylvania and toward nonunionized mines in the southern Appalachian region during the 1920s.[32] At the same time, new coal mines were opened in western states such as Colorado, Utah, and Wyoming, where union activity was restricted. In many of these western mines, companies also began turning to strip mining techniques. In these new surface mines, laborers were employed as operators of large earth-moving equipment. Because these workers faced lower levels of danger, and they could be replaced in the case of a strike, they tended to be less militant than their underground counterparts (Harding 1946: 47). As a result of this strategy of relocation, companies were able to partially counteract the militancy that had become endemic in more established centers of coal production.

It is clear that, during the interwar period, coal companies in Western Europe and North America made strenuous efforts to contain labor unrest, increase productivity, and improve their competitive position with respect to oil companies. At the same time, coal miners understandably fought to protect their autonomy and health within their workplaces. In many ways, then, the largest coal sectors in the world were caught in a grim, drawn-out conflict over issues of profitability and worker safety that were not easily resolved. The companies did have some success in imposing their solutions during the economic crisis of the early 1930s. By the late

1930s, however, coal miners across Europe and North America were in the process of reconstituting their union organizations. The impact of the Second World War would demonstrate in very dramatic terms that coal miners were able to devise new responses to the company strategies of the 1920s and 1930s. The resulting intensification of labor/management conflict in coal would open further opportunities for oil to consolidate its position as the world's primary source of commercial energy.

The Impact of the Second World War on Global Energy Industries

The build-up to the Second World War differed in many respects from that of the First, especially with regard to efforts by major governments to gain access to energy resources. Whereas in the first conflict state authorities had made few preparations to ensure they had adequate supplies of coal and oil, these concerns became absolutely central to state authorities in the decade that preceded the Second World War.

Three major powers—Germany, Italy, and Japan—had been essentially locked out of direct participation in the international oil system in the interwar period. The various commercial agreements that restricted access to Middle Eastern and East Indies oil favored companies from Britain, France, and the United States, while providing no mechanism for integrating new participants in a peaceful manner. Instead, the dominant oil companies were able to use their privileged access to prolific oil reserves to establish virtual monopoly control over the domestic oil sectors of the excluded countries. As a result, Germany, Italy, and Japan had become almost completely dependent on purchases of oil from foreign-controlled oil companies in the 1930s.

As geopolitical tensions between the major powers began to escalate in the 1930s, state authorities in the excluded countries became increasingly convinced that they had to establish their own, independent access to the energy resource that was now a central determinant of military strength. As Germany, Italy, and Japan embarked on major programs of military expansion, they were forced to directly confront the bottlenecks they faced with regard to gaining access to critical oil supplies.

In the early 1930s Germany was one of the world's largest producers of coal, while both Italy and Japan struggled to mine enough coal to satisfy the demand generated by their growing domestic industries. However, none of these countries had independent access to secure sources of oil. Even though German and Italian capital had been invested in Middle Eastern oil sectors in the pre–First World War period, participation by companies from these nations in the Red Line and Achnacarry agreements was prohibited after the First World War. The discovery of large oil reserves in Saudi Arabia in the 1930s led to renewed efforts by German, Italian, and Japanese companies to obtain drilling rights, but these efforts were unsuccessful. After 1937 Germany turned to South American countries such as Bolivia and Mexico in its quest for oil. However, these countries were pressured by the United States to limit their exports to the rising military state.[33]

Oil was particularly important to the Hitler regime, which had developed a motorized strategy of attack known as the blitzkrieg. This approach to combat relied on rapid attacks by oil-powered tanks, trucks, and aircraft. The Nazi regime therefore became acutely conscious of its dependence on foreign-controlled oil supplies. One strategy that was pursued in response to this dependence was to try to increase the synthetic production of oil from domestic coal resources. German engineering expertise was directed toward synthesizing oil from coal, so that by the late 1930s the country had become the world's largest producer of synthetic oil. Despite these costly efforts, however, Germany still imported more than two thirds of its oil in the late 1930s.[34] As a result, the nation fell back on the strategy of rapidly invading foreign territories to seize the oil reserves that were vital to its war machines. Indeed, Germany's invasion of the Soviet Union was prompted in large part by its desire to establish control over oil reserves in the Caucasian region.

Japanese military policy was driven by similar motivations. Although oil supplied less than 10 percent of Japan's total energy requirements in the 1930s, it had become critically important to the Japanese Navy. The Japanese leadership had long attempted to diversify its sources of oil. The invasions of Machuria and Korea were motivated in part by the desire to develop Japanese-controlled sources of the resource. These efforts largely failed, though, and into the late 1930s Japan was still purchasing 80 percent of its oil from the United States and the rest from the East Indies.[35] As a result, Japanese naval commanders became convinced that the only solution was to take control of the East Indies, which was controlled by the Dutch (Yasuba 1996). The need by Japan to gain access to vital oil reserves drew the nation into an increasingly hostile interaction with the United States, which eventually culminated in the war in the Pacific.

The build-up to the Second World War clearly reveals the dangers inherent in any energy system that locks major powers out of peaceful access to an era's key resources. The course of the war, meanwhile, highlights the extent to which military strategy had come to be shaped by energy concerns. The coal mines of the Saar, the Ruhr, Silesia, and the Donbass were again targeted for invasion, for instance. Meanwhile, the oil fields of Romania, the Soviet Union, the East Indies, and the Middle East were repeatedly targeted by contending forces. The ultimate outcome of the war was in many ways determined by the oil resources that each coalition had at its disposal.

In this respect, the Allies had a strong advantage. The United States was still the source of over two thirds of the world's oil, and so American oil companies were able to supply virtually all the oil that was needed by the Allies from 1941 to 1945. Although German submarine attacks took their toll on transatlantic shipments, the volume of oil emerging from North and South America was so large that these losses did not really hinder the military advance of the Allied campaigns in Europe or the Pacific. For their part, Germany and Japan were able to temporarily capture oil fields in Romania, the Soviet Union, and the East Indies. However, these facilities were severely damaged by retreating armies and were not able to supply much oil. By the middle of the war, both Germany and Japan were facing growing shortages of the resource. It became difficult to supply their dispersed military forces, which progressively weakened their fighting capacity and hastened their retreat back to their oil-deprived homeland regions.[36]

While coal sectors experienced severe devastation in Europe and Japan during the war, energy industries in the United States underwent rapid growth. Electrical systems were again expanded during the war, while government-supported manufacturing programs increased the mass production of new kinds of aircraft, motor vehicles, and ships. Meanwhile, a massive construction project was undertaken by the U.S. Army and private companies, which succeeded in building two pipelines from Texas to the East Coast. Although initially used to move oil, these pipelines were soon transformed into natural gas distribution systems that facilitated a postwar boom in gas consumption (Johnson 1967). In what proved to be the most important energy-related transformation of the twentieth century, scientists set off nuclear reactions that resulted in the release of unparalleled amounts of energy. Although first used in weapons, nuclear power would soon become the most intensively funded, government-supported civilian energy system ever deployed.

The Second World War clearly stimulated growth in a variety of energy industries. However, war again had a brutal impact on key coal mining regions in Europe. The pressures of the war also generated rising social tensions within coal industries across the world. As the military conflict drew to a close, these tensions exploded in major waves of labor unrest that, unintentionally, helped accelerate a shift toward greater reliance on oil and other resources in the postwar era.

Escalating Social Conflict in Coal

At the end of the Second World War, coal industries went through a convulsion of labor unrest that surpassed the spikes of conflict that had been generated by the First World War. In combination with the material destruction that had occurred in important coal fields and the annihilation of an entire generation of European miners, these strikes threw the international coal system into even more serious disarray.

Of all core nations, the United States witnessed the most serious labor militancy during and after the First World War. Even before the United States joined the war, the 1941 coal strike was already hampering efforts to supply allies with resources. A much larger coal strike in 1943 then caused the federal government to take control of the industry. Miners were forced back to work, though they were granted large wage concessions in most cases. Then, immediately after the war, U.S. coal miners again went on strike. Because the strike of 1945–46 began to hinder reconstruction efforts in Western Europe and Japan, the federal government pressed coal companies to raise the wages of unionized miners. As a result of repeated successes of strikes during the period 1941–46, coal miners rose to the top of the wage hierarchy in U.S. industry. At the same time, however, the market position of coal was further weakened.[37]

Similar waves of unrest swept through other established coal mining countries as well. In Britain, strikes in 1941, 1942, and 1944 forced the government to authorize substantial wage increases. Massive strikes hit the French coal sector in 1946–48, and the government was forced to grant coal miners large wage increases. In Western Germany, meanwhile, intense labor problems developed after the war, which led to the granting of widespread wage concessions. In both Canada and Australia, coal miners were major participants in strike waves during the period 1943–44 in their respective countries. Japanese coal miners also carried out large strikes during the years 1948–53. In each of these countries, coal miners forced companies and governments to improve working conditions and raise pay rates after the war.[38]

Interestingly, this postwar wave of labor/management conflict was not restricted to coal mines in developed regions of the world. For the first time, large strikes were also registered in mines throughout the developing world. Indian and Malaysian coal mines, for instance, were hit by their first large strikes in the years 1946–48.[39] Massive strikes also spread through

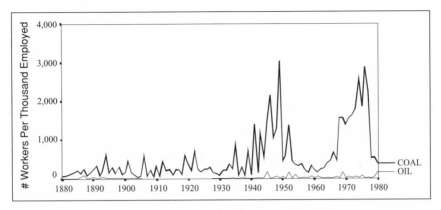

FIGURE 4.3. Strikes in U.S. Coal and Oil Industries, 1880–1980
Sources: U.S. Bureau of Labor Statistics (see Appendix C).

coal sectors in South Africa, Rhodesia, and Nigeria between 1946 and 1949.[40]

While rising labor militancy in coal was putting pressure on coal opera-tors, all available evidence suggests that oil industries were free from such difficulties. Figure 4.3, for instance, presents official strike data for U.S. coal and oil industries during the period 1880–1980, revealing again that coal dwarfed oil in strike-proneness in the one country with sizeable coal and oil sectors and comparable strike statistics.[41]

The difficulty with relying solely on official strike statistics, however, is that few countries have data on industry-level strike rates for the pre–First World War period. To map out global patterns in energy strikes, it is therefore necessary to turn to alternative sources of data on labor militancy.

One important alternative source of information on labor unrest can be found in the data set assembled by the World Labor Research Group (WLG) (see Silver 2003). The WLG data set consists of a comprehensive collection of mentions of labor unrest contained in two newspapers, the *London Times* and the *New York Times*, for the period 1905–90.[42] Detailed information is recorded for each mention, including the industry affected. Because these newspapers had relatively complete world coverage for this entire time period, the collection of labor mentions is much more global in scope than official strike indexes. Also, because newspapers are attuned to nonroutine, transformative events, the mentions often do a better job than official strike indexes of picking up on truly significant waves of labor unrest. The more than 80,000 mentions of labor unrest in the WLG data set therefore provide a unique opportunity to explore significant, world-scale trends in labor/capital conflict.

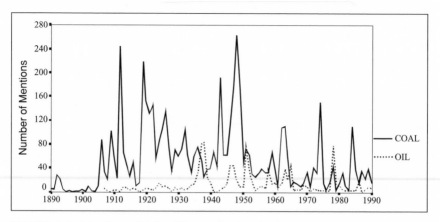

FIGURE 4.4. Mentions of Labor Unrest, Coal and Oil, 1890–1990
Sources: World Labor Group Database (see Appendix C).

An analysis of the coal and oil mentions contained in the WLG data set reveals again that the world coal system was much more prone to labor militancy than the international oil system. As shown in Figure 4.4, coal consistently surpassed oil in number of mentions during the first era of transition, except during the late 1930s when labor unrest in the Mexican oil industry captured the attention of international observers. The pattern of mentions for coal also matches the picture that emerges from the historical literature, with rising labor militancy registered in the late 1880s and early 1890s, a much larger wave prior to and after the First World War, and the largest wave in the aftermath of the Second World War. Even during the employer offensives of the late 1920s and early 1930s, coal miners continued to register high levels of militancy according to the WLG data set.

The persistent militancy of coal miners recorded in historical literatures, official strike series, and the WLG data set explains one important puzzle arising from an analysis of energy shifts. The puzzle is this: given the existence of influential companies and large labor forces in coal, why were coal interests not more successful in demanding protection from national governments in the face of the onslaught of oil? Coal companies and laborers certainly tried to win preferential treatment in such countries as Britain and Germany during the early 1920s, and such efforts appear to have temporarily slowed the French government's quest for foreign oil reserves. However, national governments could not afford to remain hostage to periodic strike waves in coal. The decline of institutional support for coal industries in the post–First World War period clearly reflected the determination of various governments to reign in the power of organized labor in coal throughout the core.[43] In sum, the lack of concerted state protection of established coal sectors in most core countries clearly reflected the need to diversify away from reliance on strike-prone coal sectors and toward oil sectors, where it was perceived that labor could be more easily controlled.

By the end of the Second World War, a variety of factors were clearly favoring a continued shift away from coal and toward new energy systems. As we shall see, a convergence of political, commercial, and social dynamics generated an expansion in world petroleum production that was unparalleled in human history. At the same time, natural gas and nuclear power industries also began periods of significant growth. Once again, the global energy system would demonstrate a remarkable capacity to set off on new trajectories of development that would bring both promise and peril to communities across the world.

5 The Rise of Oil, Natural Gas, and Nuclear Power

At the end of the Second World War, the global energy system began a new period of growth that was more rapid than any ever before witnessed in human history. In fact, between 1946 and 1973 the world consumed more commercial energy than had been used in the entire period from 1800 to 1945. While the world consumed around 53 billion tons of oil equivalent of energy in the 1800–1945 period, over 84 billion tons of oil equivalent were used in the twenty-seven years that followed the war.

This postwar boom involved the exploitation of a variety of energy resources. These years witnessed the emergence of large-scale natural gas and nuclear power industries, as well as a recovery of world coal production. The most dynamic energy industry of all, however, was oil. Indeed, world oil production grew by more than 700 percent in the period 1946–73.

As this chapter demonstrates, the meteoric rise of the international oil system was made possible by the convergence of specific geopolitical, commercial, and social factors. Under the protective umbrella of a new U.S. hegemonic order, major corporations were able to create a fundamentally new kind of energy regime. For the first time in history, the advanced industrial world ceased being self-sufficient in energy. Instead, the global South came to supply the most important energy resources being consumed in the global North. When social unrest periodically emerged to threaten these international transfers of energy resources, overt and covert force were used to reimpose order. Although this flow of resources would eventually prove to be vulnerable to interruption, for decades this international oil system provided a huge volume of inexpensive energy to consumers in the advanced industrial world.

Oil and U.S. Hegemony

The massive expansion in the international oil system that occurred in the period 1946–73 was driven by the same geopolitical, commercial, and social dynamics that had caused the international coal system to flourish a century earlier. We have seen how the expansion of the international coal system was associated with an intensified process of colonial conquest that grew out of dynamics of intra-European geopolitical rivalries. Similarly, the postwar growth of the international oil system was linked to a process of imperialist expansion fueled by growing tensions between the United States and the Soviet Union.

Overlaying these geopolitical rivalries in each case was a hegemonic state that established enough order on an international level that massive new investments in energy and transport infrastructures could be undertaken by private companies. In the earlier case of coal, the British government supported companies in their efforts to develop mining projects in colonial regions. Similarly, in the postwar era the U.S. government exerted its influence in the Middle East and Latin America to ensure that western companies were able to gain access to petroleum reserves and begin the process of transporting them to markets in core nations.

Finally, the international expansion of each new energy system was fostered by government and corporate successes in containing social tensions in key zones of production. Just as European and North American governments assisted in the process of holding labor militancy at bay in coal mines during the latter part of the nineteenth century, these same governments worked with multinational energy companies to contain nationalist unrest in key regions of oil production in the two decades following the Second World War. In each case, government policies also shifted toward strong support for increasing reliance on the era's new energy resource. As we now turn to an analysis of the postwar period, we will seen how these intersecting geopolitical, commercial, and social dynamics fostered the global expansion of the international oil system.

Let us begin by examining how dynamics of geopolitical rivalry between the United States and the Soviet Union fueled growth in oil production and consumption in the regions that each superpower came to dominate. From the perspective of American strategists, the building of a western alliance led by the United States came to be seen as dependent on stable economic growth in Western Europe and Japan. Such economic expansion, in turn, was seen to require the provision of cheap oil to alliance partners. Soviet authorities followed a similar line of analysis, which suggested that a strengthened eastern bloc would similarly require access to low-cost oil reserves.[1]

In the immediate postwar period, the Soviet Union faced a much easier task of attaining these strategic oil objectives than did the United States. Huge oil reserves were known to exist in the Soviet Union and in states such as Romania that were solidly under Soviet control. Soviet planners and their local allies were therefore able to utilize command-style economic policies to construct the infrastructure, impose the pricing agreements, and coordinate the movement of supplies from oil wells in secure regions to designated consumers. Moreover, the authoritarian states that spread across the eastern hemisphere after the war were able to impose rigid controls on labor costs and social unrest. This combination of circumstances permitted the rapid construction of oil pipelines and shipping terminals in the Soviet Union and its allies in the early postwar period, which then allowed massive quantities of low-cost oil to be distributed to socialist states over the next three decades.[2]

In contrast, the United States confronted many more challenges in its efforts to construct a system for the distribution of oil resources in its own sphere of influence. To begin with, while U.S. domestic reserves could still satisfy burgeoning American demand, they were not prolific enough to fuel economic reconstruction in Western Europe and Japan. Even in the midst of the war, American military planners recognized that massive oil reserves in the Middle East and Latin America would also have to be harnessed for this project to succeed. However, unlike the compliant political context that existed in the Soviet sphere of the 1950s, U.S. policy makers faced a complex terrain of independent and frequently contentious states who either had oil or else wanted access to it. In what would become a hallmark of U.S. hegemony, American strategists and corporate executives therefore made use of a combination of political power and market dynamics to foster the growth of an international oil system that could supply expanding western economies with oil.

The first steps in constructing this oil system were overtly military in nature. During the final years of the Second World War, beneath the formal civilities of a cooperative wartime effort with Britain and France, the United States engaged in a systematic campaign to increase its access to foreign oil supplies. In 1943 both Iran and Saudi Arabia were declared eligible for lend-lease aid, and by the end of the war tens of thousands of U.S. soldiers had been deployed to the region. By 1944, U.S. companies controlled around 40 percent of known Middle East oil reserves. Moreover, American forces in the Pacific prevented the Dutch from restoring control over Indonesian oil fields. Strong diplomatic pressure was also exerted to reduce European influence in Latin American oil industries as well.[3]

The application of direct military and diplomatic influence ensured that U.S. corporations got access to key oil fields in the immediate postwar period. But within a few years, this overt exertion of American power was replaced by more subtle forms of U.S. influence. Multinational oil companies were given the central responsibility of constructing and operating the international oil system. These corporations were assisted in their efforts to develop new oil-based infrastructures by the World Bank, which was able to mobilize investments under the rubric of national development. They were also supported by the financial protections provided by the International Monetary Fund and the pressure exerted by the General Agreement on Tariffs and Trade on peripheral countries for openness to foreign direct investments.[4]

With these support structures in place, the corporations that had come to dominate the international oil system were able to negotiate extremely favorable deals with elites in oil-rich countries. In most years, and in most places, this postwar oil regime operated as a private, market-based system that required little direct intervention from the U.S. government or its allies. In this respect, the international oil system differed significantly from the earlier coal system that had been constructed on the basis of much more overt forms of colonial intervention.

It also differed in another important way. The hegemonic order constructed by Britain in the nineteenth century fostered the movement of coal resources out of the United Kingdom to key colonial outposts, and it supported the construction of new mines in the colonies to sustain local consumption. British hegemony therefore promoted increased coal consumption in the periphery, because heavy transport systems were constructed in the colonies for military, economic, and social purposes. In contrast, the hegemonic order built by the United States in the twentieth century fostered the extraction of oil resources from the periphery for consumption in core economies. Though some elites in peripheral nations got access to a portion of this petroleum, the market dynamics underlying the international oil system ensured that the bulk of the resource was shipped to core consumers who could pay for it. One twist, then, is that the more coercive era of British hegemony promoted a more equitable distribution of coal, while the more market-driven form of U.S. hegemony fostered an extremely unequal distribution of oil.

Although it was based on less overtly coercive forms of control than were present in the nineteenth-century coal system, the U.S.-centered international oil system was tremendously successful in moving huge volumes of the resource from the periphery to the core in the period 1946–73. This international transfer of oil, in fact, became a central component of U.S. hegemony. In addition to its growing military superiority and competitive economic advantages, it was crucial that the United States be able to build legitimacy for its hegemonic project by fostering American-style forms of mass consumption in allied states.[5] Just as the British had pointed to the construction of railways and shipping lines to justify their colonial rule, so too did American statesmen tout the postwar expansion of automobile ownership and civilian air travel in Western Europe and Japan. The fact that these new forms of consumption were ostensibly driven by market dynamics and seemed to enhance individual mobility and freedom gave added legitimacy to this image of an oil-based society built on the American model.

Of course, there were contradictions embedded in this oil-based lifestyle. As would become clear decades later, the freedom that many people in core countries thought they had won by purchasing a car became constrained by traffic gridlock and heavy auto-related charges. Moreover, the diffuse urban landscapes fostered by automobiles generated a host of environmental and health threats. But these problems would not fully manifest themselves in core countries for some time.

Of more immediate concern for the stability of the international oil system were the contradictions faced by people in peripheral societies. In Latin America and the Middle East, oil markets functioned to extract resources at the lowest possible cost—and with few enduringly positive local impacts. In fact, while oil may have fostered certain forms of freedom in core countries, it was associated with the emergence of authoritarian governments and military interventions in the peripheral countries. As a result, even during the period 1946–73, when the international oil system operated in its most stable form, there were significant instances in which political and social unrest threatened the system. The responses that these outbreaks of unrest elicited from core states revealed that this energy system was not solely constructed on laissez-faire market principles. While the United States was more subtle in its use of force than the United Kingdom had been, American and European strategists proved that they were willing to use overt force to contain threats to the oil system in the decades following the Second World War.

A moderate challenge came during the Second World War, when Venezuelan officials demanded that oil corporations agree to a fifty-fifty profit split. The multinationals operating in the country initially refused to accede to the demand. They contacted the U.S. State Department to request that diplomatic pressure be applied on the Venezuelan government to withdraw the demand. However, American officials were principally concerned with cementing Venezuelan allegiance during the war and they instead urged the companies to agree to the concessionary revision. By 1948 Venezuela had won better terms, and similar contract revisions spread throughout the Middle East in the following years.[6]

While reforms in petroleum contracts modestly reduced the profit margins enjoyed by the multinationals, these commercial adjustments did not fundamentally challenge the normal course of business in foreign petroleum industries. However, nationalist regimes emerged in the 1950s in Iran, Egypt, and Iraq that posed a more serious threat to the international oil system. The responses revealed that, under all the subtleties of American hegemony, there existed a willingness to use overt power to protect the interests of key core economies in the face of peripheral unrest.

The first radical challenge to the post-war international oil system came in Iran during the early 1950s. Widespread popular discontent at the fact that the British owned virtually all of Iran's oil reserves had been building since the 1920s. In 1951 the politician who most insistently called for the nationalization of the industry, Muhammad Mossadegh, was democratically elected as Prime Minister. Shortly after taking office, Mossadegh began the process of nationalizing the Anglo-Iranian Oil Company. Though the nationalization policy was tremendously popular within Iran, it was unacceptable to the oil multinationals and their host countries. In 1953 American and British special forces orchestrated the overthrow of Mossadegh and the installation of an authoritarian government headed by the pro-Western Shah Pahlavi, who allowed foreign companies back into Iran under extremely favorable conditions. Following the Iranian crisis, the United States increased its level of military assistance to the Shah and to the governments of Israel and Saudi Arabia.[7]

Soon after the Iranian crisis was contained, another emerged in Egypt. Though Egypt did not hold important quantities of oil, the Suez Canal was located within its borders. Owned and operated by Europeans, the canal had become a crucial conduit for oil resources flowing from the Middle East to Western Europe. However, relatively little money was paid to the Egyptian government for use of the canal. The fact that such an important piece of Egyptian territory was owned by foreigners became a matter of increasing concern within the country. Finally, President Gamal Nasser nationalized the canal in 1956, to great acclaim in Egypt and throughout the Middle East.

Although Nasser promised to compensate the owners for their property, the reaction was swift. British, French, and Israeli forces were deployed to Egypt in an attempt to forcefully retake the Canal. This military response was severely criticized by most regimes in the region and by the United States and the Soviet Union. Strikes and riots spread throughout the entire Middle East in the aftermath of the invasion. In Syria, Saudi Arabia, Bahrain, Kuwait, Iran, and Iraq, oil workers carried out violent protests while numerous cases of sabotage against European pipelines also occurred. In the face of strong Egyptian military resistance, U.S. anger, and this region-wide wave of protest, the military effort to retake the Suez Canal was abandoned. Because of the collapse of the invasion, the governments of Britain and France suffered severe blows to their prestige while the United States and the Soviet Union became the most influential foreign powers operating in the region.[8]

Although the United States had refused to support the British and French in Egypt, it was soon made clear that American military power would occasionally be used to contain unrest in the region. In January 1957 the U.S. Congress approved President Eisenhower's request for authorization to dispense economic and military aid to allied states in the Middle East if they were threatened. These new presidential powers were utilized a year later, when a pro-Western monarchy in Iraq was overthrown by nationalist generals who received support from the Soviet Union. When unrest also spread to Jordan and Lebanon, Eisenhower responded by deploying 14,000 troops to Beirut to stabilize those regimes. Over the following decade, the Kennedy and Johnson administrations radically increased U.S. economic and military assistance to Israel, Iran, and Saudi Arabia, while the Soviet Union supplied Egypt, Iraq, and Syria with similar assistance. Although these rising Cold War tensions served to dampen local crises during the 1960s, the militarization that occurred in the region would set the stage for much greater conflagrations in future decades.[9]

The 1960s were a period of remarkable stability and growth in the international oil system. Social and political threats in key peripheral countries were largely contained by the delicate balance of power that was established between the United States and the Soviet Union. At the same time, massive discoveries of new oil resources provided a broad terrain of operation for the multinationals. Indeed, from a commercial perspective the main challenge involved coping with the chronic overabundance of oil that resulted from continuous discoveries. Recall that, in the previous century, profit rates in many coal sectors had been decimated by the intense competition that followed the discovery of abundant mineral reserves. By contrast, a much smaller number of multinational petroleum firms succeeded in containing price competition during the period of oversupply that characterized the 1960s.

A key factor that allowed price competition to be contained during this period was the willingness of the largest oil multinationals—such as British Oil and Royal Dutch Shell—to allow new firms to participate in Middle East oil development. This magnanimity was partly motivated by pressure from the U.S. government, which demanded increased access for American companies. It was also part of a corporate strategy to contain competitive dynamics in the Middle East. A central worry was that new American entrants like Standard of New Jersey, Standard of New York, and Gulf Oil would try striking independent deals with governments in the region. This could set off a new escalation in demands for concessionary revisions, which would have hurt the interests of all companies. To avoid this possible outcome, the long-established companies allowed newcomers to enter into joint ventures to extract newly discovered oil under preexisting contracts.[10] This strategy of corporate accommodation absorbed the huge discoveries of oil made in Kuwait, Saudi Arabia, and other Middle Eastern kingdoms without provoking ruinous competition or a wave of nationalizations. The success of this approach is illustrated by the fact that, in 1972, eight multinational petroleum companies controlled over 75 percent of the noncommunist world's oil reserves, including over 90 percent of Middle East production.

Throughout the 1960s, the international oil system functioned remarkably well. The geopolitical tensions that developed between the United States and the Soviet Union prompted these nations to support increased oil production and consumption in their zones of influence, while the militarization programs they sponsored in the Middle East stabilized local regimes and helped contain serious social crises in that key region. Meanwhile, the multinational oil companies set up a system for absorbing new reserves and companies that succeeded in preventing the emergence of cutthroat forms of competition or serious nationalization threats.

Given this context of geopolitical, commercial, and social stability, a massive wave of private and public investments flowed into new petroleum projects throughout the world. It is estimated that between 1955 and 1970, private investments in foreign oil exploration, extraction, and marketing projects totaled some $215 billion. There was also a dramatic expansion in oil pipelines, transoceanic tanker tonnage, and other oil-related infrastructure.[11] Multilateral lending agencies such as the World Bank played a central role in developing the transportation infrastructure that was needed to move oil resources from zones of production in the periphery to key markets in the core. The reduction in transport costs achieved with this infrastructural expansion allowed the low-cost oil extracted from peripheral wells to reach North American, European, and Japanese markets without incurring significant price markups; as a result, the average price of oil finally became cheaper than that of coal in most western cities in the 1950s. This price advantage then led to a burgeoning consumption of oil.

The rapidity with which oil overtook coal as the world's principal source of commercial energy during the period 1946–73 is astonishing. The percentage of total world energy provided by coal was cut in half, falling from 62 to 28 percent in less than thirty years, while oil's share rose from 27 to 49 percent. This represents a remarkable increase in the volume of oil that was extracted and consumed. In 1946 the world consumed around 1.3 billion metric tons of oil, while in 1973 annual consumption had more than quadrupled to 5.5 billion tons. Whereas coal had powered the industrial revolutions of the eighteenth and nineteenth centuries, and even maintained its predominance during the first half of the twentieth century, it was rapidly overtaken by a flood of low-cost oil in the post–Second World War era. Once again, the global energy system demonstrated an amazing ability to alter its material foundations in a space of less than thirty years.

The Expansion of Natural Gas

In addition to the massive growth of the oil system, the postwar era saw the consolidation of yet another important international energy industry—that of natural gas. Again, the emergence of this new energy resource did not reflect any material scarcity. Instead, the expansion of natural gas was driven by many of the same advantageous political, commercial, and social factors that stimulated the postwar growth of oil.

As with coal and oil, natural gas had been used in small quantities by humans for thousands of years before it was developed into a major source of energy. The first documented use of the resource came in Persia between 6000 and 2000 B.C., where seeps from the ground were ignited for fire-worshiping ceremonies. By around 200 B.C., Chinese engineers were drilling into the earth to get access to gas to burn in rock salt industries. Centuries later, small quantities of gas were used in Western Europe and North America to supplement coal gas in street lights. But the diffuse characteristics of the resource made it more difficult than coal or oil to transport, and so natural gas was largely ignored as a source of commercial energy until the beginning of the twentieth century.[12]

It was in the United States where the first large-scale commercial gas industry emerged. This development reflected both geological and social factors. Most natural gas at the time was encountered by oil companies, since the two resources are often embedded in the same geological formations. So, along with major discoveries of oil in the United States and other regions came discoveries of significant reserves of gas. Given the complexities involved in transporting the gas, though, it was normally flared away or else injected back into the earth to increase oil extraction. These practices continued through the 1920s, when oil and utility companies began experimenting with pipeline systems designed to deliver the gas to consumers. These efforts were aided by the fact that many sites of gas extraction in the United States were near cities, and so infrastructural costs were manageable. In contrast, the gas discovered in Eastern Europe, Latin America, and the Middle East were far from sites of consumption, and so flaring continued for much longer in these regions.

This combination of prolific supplies and accessible markets favored the expansion of a commercial natural gas industry in the United States. This growth was further stimulated by specific private sector dynamics and by government intervention in the new industry. In terms of private sector factors, gas benefited from the fact that well-capitalized oil companies were the ones with access to the new resource. Petroleum companies were uniquely able to invest the money required to develop pipeline systems and build the early infrastructure required to test the viability of large-scale commercial projects. Moreover, high barriers to entry meant that only a small number of companies could get into the business. This translated into steady profit margins for those who were able to undertake gas projects.[13]

While private capital drove the early growth of the natural gas industry in the United States, the federal government soon intervened in ways that were critical to the industry's expansion. For instance, the federal government was centrally involved in accelerating the development of a nationwide pipeline system. During the Second World War, the government directed that two large pipelines be constructed from Texas to the East Coast to move oil in a secure manner. After the war these lines were converted to gas transmission, and the new resource experienced a jump in consumption. In addition, federal authorities imposed price controls on the industry in the 1950s, which spurred demand for gas in the suburbs that were growing around the nation. Finally, federal authorities passed legislation to prohibit the flaring of gas by oil companies. When a few large fines were imposed on petroleum companies in the mid-1950s, compliance with these new conservation laws improved and gas deliveries to consumers rose steadily.[14]

It is also important to note the role of social pressures in fostering the development of the natural gas industry in the United States. We have seen that a major wave of strikes swept through coal mines after the Second World War. These strikes disrupted deliveries of residential coal just at a time when new pipelines were being opened for natural gas deliveries. One unintended result of labor conflicts in coal, then, was to cause an important number of real estate developers and property owners to install natural gas heating systems in new homes. Given social unrest in coal and favorable conditions in corporate and governmental arenas, the stage was set for a shift toward increased reliance on gas in key markets.[15] The result was more than a four-fold increase in natural gas consumption in the United States in the period 1950–73.

While the United States pioneered the development of a nationally integrated natural gas industry, Western Europe also witnessed a rapid expansion of gas consumption in the postwar era. Large discoveries of gas in the Netherlands and the North Sea in the late 1950s spurred the industry. Indeed, by the 1970s the Netherlands had become the world's largest exporter of natural gas, while consumption of the resource had expanded dramatically in surrounding nations.[16]

Just as in the United States, the multinational oil companies were centrally involved in developing these new reserves of gas in Western Europe. This again favored the gas industry, because deep financial resources were available to develop the region's gas infrastructure. Also as in the United States, Western European governments used regulatory mechanisms to accelerate the expansion of the industry. These promotional efforts were driven primarily by a desire to reduce the use of coal in urban residences in order to reduce coal-related pollution that had become endemic in cities across the region. The threat of coal-fogs was most starkly illustrated by the disaster that struck London in 1952, where over four thousand residents died because of toxic air (Stone 2002). This crisis generated strong public and political pressure to shift to cleaner gas systems in urban homes, and by 1956 new legislation had restricted the use of coal in many English cities. By the next decade, similar laws had been passed in most other countries in the region. As a result of this convergence of corporate, governmental, and environmental factors, natural gas consumption in Western Europe increased by over 800 percent in the period 1950–73.

While impressive expansions in natural gas consumption were achieved in North America and Western Europe in the decades following the Second World War, growth in other regions was more modest. The Soviet Union initiated some gas projects in the Volga, North Caucasus, and Central Asian republics in the 1960s. However, the fact that the most prolific gas reserves were located far from industrial or population centers served to limit the use of the resource. Meanwhile, a modest increase in gas exports from countries like Algeria, Iran, Brunei, Libya, and Indonesia occurred in the 1960s. It is estimated, though, that most of the gas encountered in oil wells was flared off by petroleum companies. This inefficient practice was due to the fact that pressurized ships were not yet in widespread use, and so it was difficult to move gas to consumers in cities across the world.[17]

Overall, natural gas was absorbed into the global energy system in a uniquely wasteful fashion. It took decades for companies in North America and Eastern Europe to begin systematically harnessing the resource. In many regions of the developing world, tremendous volumes of the resource continue to be flared or vented today. Still, in certain areas where favorable geological and social factors converged, this new source of energy was put to good use. The decades after the Second World War saw significant growth in the proportion of gas that was harnessed for commercial use. Indeed, between 1949 and 1973 world natural gas consumption rose from 266 million tons of oil equivalent to over 1.2 billion tons of oil equivalent. Once again, the global energy system demonstrated a remarkable capacity to incorporate massive quantities of new energy resources in the space of less than thirty years.

The Expansion of Nuclear Power

If we look at trends in commercial energy growth, the most significant change that occurred in the global energy system in the decades following the Second World War was clearly the massive influx of oil and gas resources. However, if we alter our frame of analysis to take into account geopolitical and environmental considerations, it can be argued that the most important postwar energy transformation was the emergence of nuclear power. Of all the energy-related impacts of the Second World War, perhaps the most significant was the fact that the conflict spurred the development of atomic weapons. Once the bomb had been deployed, there followed a tremendous push by government officials to create civilian nuclear power industries. As a result, intertwined military and civilian nuclear industries spread across the world in the decades following the war. By the mid-1980s, nuclear power had become a significant source of electricity in North America, Western and Eastern Europe, and Japan.

As with earlier energy systems, the rise of nuclear power did not occur because of pressing resource needs. On the contrary, the foundations of a civilian atomic industry were laid during a time of historically unparalleled growth in long-established energy sectors. The abundance of coal, hydroelectric, oil, and gas resources—combined with reduced demand caused by the lingering effects of the war—drove the prices of energy commodities and services to record lows in most countries. The economics of energy were strongly against the introduction of a very expensive, technically complex, and hazardous new energy resource. Yet, in country after country, nuclear sectors expanded in the decades following the Second World War.

If there was no pressing economic reason for developing civilian nuclear power sectors, then what explains the steady growth of this new energy regime following the war? Simply put, this growth reveals in very clear terms the power of the state to promote new energy systems. Unlike with coal, oil, or hydroelectricity, nuclear power was not initially created and commercialized by private companies. Quite the opposite was true, in fact. In every country that witnessed the consolidation of a civilian nuclear power sector, state officials had to employ hard-sell policies to induce utility companies to participate in nuclear projects. This intense state support was provided because, from its inception, civilian nuclear power has had intimate connections with military power and national prestige.

The basic science leading to the harnessing of nuclear power was carried out by European scientists in the early decades of the twentieth century. Albert Einstein's proposition that all matter was energy was of crucial importance to this endeavor. Einstein's theoretical work was also supported by laboratory research carried out in England, Italy, France, Austria, and Germany by scientists like Rutherford, Fermi, the Curies, Bohr, Hahn, Meitner, and Strassman. Accumulating knowledge about radioactive elements led to the first controlled fission reaction in Germany in 1938. Although many of the scientists who pioneered this research fled that country soon afterward, Hitler's regime continued to finance research into atomic energy. By 1939 leading physicists had convinced the U.S. government to fund its own project to construct an atomic bomb. In 1942 a controlled fission reaction was created in Chicago, and on July 16, 1945, the first atomic bomb was exploded in New Mexico. Less than a month later, atomic weapons were dropped on Hiroshima and Nagasaki, Japan.[18]

The destructive power of these early atomic weapons was unprecedented in human history. It is estimated that over 150,000 Japanese citizens were killed almost instantly, while tens of thousands more died over the following decades because of lingering effects of the bombs. This lethality had an immediate impact on the community of nuclear scientists who had participated in the creation of the bomb. Many of these scientists, led by Albert Einstein, called for a world moratorium on the further development or use of nuclear technologies. Conversely, scientists such as Edward Teller argued strongly that nuclear power could be harnessed for peaceful purposes.

A powerful convergence of military planners, university researchers, and corporate executives eventually succeeded in convincing the Truman Administration and Congress that atomic energy should be incorporated into civilian power systems. Under prompting from this military-industrial complex, the U.S. Congress passed the Atomic Energy Act of 1946.[19] This legislation sanctioned the construction of civilian reactors, and it also created the Atomic Energy Commission to help promote the commercialization of the new, untested technology.

The response from domestic U.S. utilities companies was less than enthusiastic, however. To stimulate the private development of nuclear power, the Atomic Energy Commission provided myriad supports to U.S.

firms engaged in building and operating nuclear power stations, including underwriting reactor construction costs, providing free fuel for reactors, funding nuclear research and development activities, and committing the federal government to the development of nuclear waste disposal facilities.[20] The Eisenhower Administration also persuaded Congress to enact the Price-Anderson Act of 1957, which limited the liability of utilities operating nuclear reactors to a maximum of $560 million in the event of any accident.[21]

With such high levels of government support, a small group of American companies finally agreed to participate in developing a civilian nuclear power industry. Westinghouse and General Electric began designing reactors and containment facilities, while firms such as Babcock and Wilcox, Bechtel, and Combustion Engineering assisted with engineering and construction activities. Mining and oil companies also participated in the emergent industry. In 1957 the country's first full-scale civilian nuclear reactor was brought online in Shippingport, Pennsylvania. Even though the heavily subsidized electricity generated by this facility was more than ten times as expensive as that provided by conventional plants, orders for additional reactors were placed in the United States. Over the next three decades, over one hundred civilian nuclear power stations were opened in the country. By 1989, at the crest of the expansionary wave of nuclear power production in the United States, civilian reactors were generating approximately 530 billion kilowatt hours of electricity a year.[22] Strong state promotion was clearly successful, pushing forward domestic reliance on nuclear power, even in the face of significant technological and commercial hurdles barriers that existed in the United States.

In addition to these domestic efforts, the U.S. government also shifted from a policy of nuclear containment to one of international promotion. This shift followed the detonation of an atomic weapon by the Soviet Union in 1949. Once it was clear that nuclear capabilities had spread beyond U.S. control, the American government significantly expanded its efforts to transfer nuclear technologies to its allies. It provided a steady supply of enriched uranium and technical assistance to countries in Western Europe and East Asia to enhance the nuclear deterrent capabilities of key allies in the context of an intensifying Cold War.

The clearest signal that the United States had shifted to a promotional stance with respect to international nuclear power came in 1953 with President Eisenhower's "Atoms for Peace" speech at the United Nations. In his speech, Eisenhower declared that his administration was intent on fostering the international diffusion of an energy technology he claimed would solve humanity's energy difficulties by the end of the century.[23] U.S.

government promotional efforts in this arena were matched by domestic research and development programs that sprang up across Western Europe immediately following the Second World War.

The most rapid expansion in civilian nuclear sectors occurred in Britain and France, each of which had long-established indigenous research traditions in atomic science. By 1952 Britain had tested its own nuclear weapon, and by 1956 its first civilian reactor entered into operation. France was not far behind in the development of weapons and civilian nuclear power plants. Promotion of nuclear power was then accelerated by the signing of the Euratom accord in 1958, which established guidelines for the transfer of nuclear fuel and technical expertise from the United States to a series of West European countries, including West Germany. At the same time, the United States sponsored legislation permitting the development of a civilian nuclear power industry in Japan. Though both West Germany and Japan were officially prohibited from developing atomic weapons, the development of civilian nuclear power sectors turned each of these states into virtual nuclear powers. Indeed, by promoting the construction of civilian reactors in Western Europe and Japan, the United States created a powerful network of nuclear allies in the 1960s to counter a rising nuclear threat from the Soviet sphere.[24]

For its part, the Soviet Union engaged in a massive drive to develop its own nuclear capacity following the Second World War. The speed of Soviet nuclear development was remarkable. The first atomic bomb test occurred in 1949, and the first civilian nuclear power reactor came online in 1954—three years before the first American plant entered into operation.[25] Having established its own nuclear capabilities by the mid-1950s, the Soviet Union then mimicked the United States by distributing its own atomic systems to key allies in Eastern Europe and Asia. Standardized reactors, radioactive fuel, and technical assistance were provided to Czechoslovakia, East Germany, Poland, Romania, Bulgaria, and Hungary during the 1950s and 1960s.[26]

Although both American and Soviet officials claimed to be fostering civilian nuclear power development, it soon became apparent that it was extremely difficult, if not impossible, to prevent countries from deploying nuclear weapons once they had initiated civilian projects. The first case of unintended weapons proliferation occurred in China. Soviet engineers shipped a reactor to China sometime after 1955, with the intention of supporting the peaceful production of electricity in that country. As a political rift developed between the Soviet Union and China, Soviet authorities became evermore adamant that their technology not be used for weapons development. Nevertheless, in 1958 Chairman Mao announced that China would indeed build its own atomic weapon. Soon after that, all Soviet nuclear assistance was terminated. This did not stop Chinese advances, however, and in 1964 the country detonated its first nuclear device.[27]

Similar patterns occurred in India and Pakistan. In the late 1950s the United States, Britain, and Canada helped India construct a reactor for what was intended to be peaceful research purposes. Through the 1960s Indian scientists developed their expertise in nuclear physics. When the country's political leaders decided in 1972 to construct a nuclear weapon, there was little that could be done by foreign powers to halt the project. In 1974 the Indian military detonated its first nuclear bomb. For its part, Pakistan contracted with American, Canadian, and French engineers for support in building a civilian nuclear reactor. Although weapons proliferation concerns eventually caused most international assistance to be withdrawn, Pakistani scientists were able to complete the reactor and a fuel enrichment plant on their own. Throughout the 1980s and 1990s, the Pakistani government steadily built up its weapons-grade fissile materials and missile systems, and in 1998 the country detonated its first large-scale nuclear bomb. In addition, the head of Pakistan's nuclear program, Abdul Qadeer Khan, headed a network of Pakistani scientists and military personnel that sold nuclear materials and technology to Iran, North Korea, Libya, and perhaps other customers.[28]

The examples of China, India, and Pakistan demonstrate that it is extremely difficult to promote the construction of civilian nuclear power stations while preventing the proliferation of military programs. Indeed, over the last three decades a number of countries have used the development of civilian nuclear industries as a step toward constructing military atomic programs.[29] For instance, Israel is widely suspected of having built a large number of weapons after having first expanded its civilian power industry. North Korea has also claimed to have constructed atomic weapons out of materials generated in its civilian power reactors. Recent evidence has also emerged to suggest that Iran is transforming its experience with civilian nuclear power systems into a weapons program.

Although there were inherent dangers in nuclear energy, these dangers took time to manifest themselves. The first twenty years of nuclear power development were in fact characterized by a remarkable growth in electrical output across North America, Europe, and Japan. In 1979, for instance, the world's civilian nuclear power stations together generated over 620 billion kilowatt hours of electricity. The example of nuclear power, therefore, demonstrates the capacity of state authorities to push energy technologies forward in the face of significant commercial resistance. Unfortunately, it would soon become clear that these massive investments in nuclear industries created enduring security and environmental problems across the world.

The Postwar Emergence of International Energy Inequalities

The post–Second World War era clearly witnessed important developments in a whole range of energy systems. In both the U.S. and the Soviet spheres of influence, specific social groups gained privileged access to new oil, gas, and electricity resources. Industry was stimulated in these regions, and significant improvements in basic standards of living were achieved. However, this process of development in core and semiperipheral regions masked an intensifying degree of inequality in global patterns of energy consumption. In the American sphere of influence in particular, market forces functioned to exacerbate divisions between those who got access to modern forms of energy and those who were forced to continue relying on traditional biomass resources.

If we look at patterns of energy production and consumption by world region in the period after the Second World War, a new problematic feature of the global energy system is revealed. The nations of the global South have been transferring modern energy resources to wealthier nations at an increasing rate in recent decades. As shown in Figure 5.1, through the end of the Second World War the developed world was almost totally self-sufficient in energy. Since then, however, nations of the global South have been transferring energy resources to nations in the global North at a steady rate.

A number of oil-exporting countries have achieved impressive levels of economic growth on the basis of this trade. However, its main effect has been to intensify long-standing global inequalities in levels of energy consumption. As indicated in Figure 5.2, throughout the modern period core states have attained much higher levels of per capita commercial energy consumption than their semiperipheral or peripheral counterparts. While there was a slight closing of the gap between core and semiperipheral regions during the 1970s, by the mid-1980s long-term patterns of intensifying inequality had reasserted themselves.

If we focus our attention on the post–Second World War period and examine world regions in more detail, we again see enduring patterns of inequality. Figure 5.3 shows that North America (the United States and Canada) has persistently outstripped all other regions in terms of

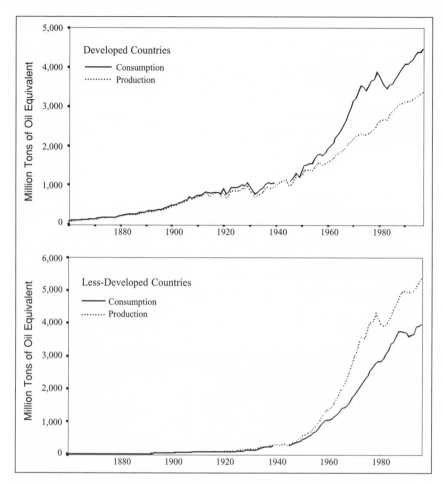

FIGURE 5.1. Commercial Energy Production and Consumption, 1860–2000
Sources: See Appendix A.

commercial energy consumption. After seeing substantial gains in the three and a half decades following the Second World War, meanwhile, countries in Eastern Europe have undergone a significant decline in consumption. Western Europe, which saw a slight pause following the shocks of the 1970s, has reasserted moderate growth. The Pacific region, which includes Japan, East Asia, and Australia, has seen steady growth. Africa and Asia, meanwhile, have seen little increase in per capita consumption of commercial energy since the 1970s.

Turning to a more focused analysis of the present situation, we again find that countries exhibit very divergent patterns of energy consumption. As shown in Figure 5.4, the average citizen in the United States consumes

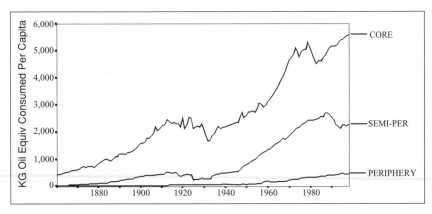

FIGURE 5.2. Per Capita Commercial Energy Consumption, 1860–2000
Sources: See Appendix A.

five times more energy than the world average, ten times more than a typical person in China, and over thirty times more than a resident of India. Even in such major oil-exporting nations as Venezuela and Iran, per capita consumption of commercial energy resources is less than one half and one quarter of the U.S. average, respectively. A starker illustration of these inequalities is captured in the estimation that around 40 percent of the world's population—over two billion people—still has no regular access to commercial energy products in their homes (World Energy Council 2000).

It must also be observed that these unequal patterns of consumption show little sign of easing. This can be demonstrated through a quintile-based analysis. As shown in Figure 5.5, in 2000 the top quintile

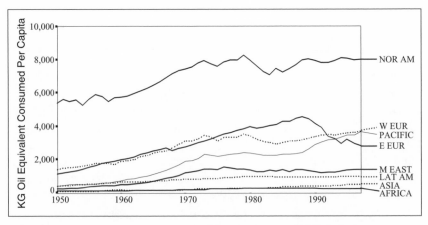

FIGURE 5.3. Per Capita Commercial Energy Consumption, 1950–2000
Sources: See Appendix A.

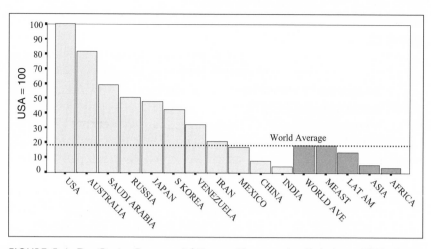

FIGURE 5.4. Per Capita Commercial Energy Consumption Relative to USA, 2000
Sources: See Appendix A.

(containing the wealthiest 20 percent of the world's population) consumed about 68 percent of the world's commercial energy, while the lowest quintile consumed under 2 percent of these resources. Moreover, the overall distribution has remained fundamentally unaltered in the post-1960 period. While consumers in wealthier regions of the world greatly expanded their consumption of oil and electricity after the Second World War, a disturbing fact began to be revealed by improving international energy statistics. Specifically, surveys increasingly documented that large

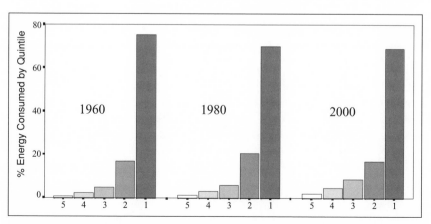

FIGURE 5.5. World Commercial Energy Consumptin by Quintiles
Sources: See Appendix A.

numbers of people in the developing world still had no access to modern energy systems.

A postwar study by the United Nations, for instance, estimated that preindustrial fuels such as firewood, straw, and dung provided over 45 percent of Latin America's energy, over 55 percent of Africa's energy, and over 65 percent of Asia's energy. Studies in the 1970s again documented major inequalities, revealing, for instance, that 90 percent of Indian citizens relied on traditional fuels for their cooking, heating, and lighting requirements.[30] The extent to which consumers in developed countries had come to monopolize the world's most desirable energy resources and technologies—leaving wide sectors of the developing world to subsist on ever-scarcer traditional fuels—would lend added legitimacy to demands for redistribution of the wealth and benefits generated by world energy industries. One of the central challenges facing the world community in this century will be to begin to alter these intensifying patterns of inequality in the global energy system.

6 The Second Period of Crisis

hile the two and a half decades following the end of the Second World War were years of unparalleled growth in world energy industries, the decade of the 1970s ushered in a new era of crisis. Turbulence rippled through the international oil system, while accidents in nuclear power plants brought to light dangers in this emerging energy technology. Evidence also began to mount that the pollution generated by conventional energy systems was having detrimental impacts on the environment, and so resistance to expansions in energy projects began to mount. By the end of the twentieth century, it had become clear that the global energy system was caught in a sustained crisis that showed little sign of relenting.

To comprehend the enduring nature of this crisis in the global energy system, we have to turn to an analysis of the world historical dynamics that have been in operation in the contemporary period. This chapter begins by examining why the geopolitical, commercial, and social factors that had fostered steady growth shifted in more chaotic directions. As we will see, one of the most intriguing effects of this chaos was that core states began supporting more efficient and novel energy systems in the 1970s. In a surprisingly short time, consumption rates were slowed and new energy systems were deployed. However, by the 1980s support for this new energy shift waned, and the world-economy settled back again into uneasy reliance on conventional energy resources. By arriving at an analysis of the systemic dynamics that first encouraged and then undermined the expansion of new energy technologies, we will be in a better position to map out both the possibilities and challenges facing a more far-reaching shift to new energy systems in the twenty-first century.

A World Historical Interpretation of the First Oil Shock

Although many analysts treat the first oil shock as a largely unpredictable event, this crisis can in fact be seen as having been fostered by a convergence of social, commercial, and geopolitical tensions that occurred in the late 1960s. Of course, it is never possible to forecast precisely how and when crises will manifest themselves. However, an examination of world-systemic dynamics can shed light on when major periods of disorder, such as that of the early 1970s, are likely to sweep through established energy industries.

As earlier chapters have shown, systemic dynamics intersected in the late nineteenth century in ways that generated a crisis in the international coal system. Rising social tensions in coal mines disrupted operations and drove production costs higher. At the same time intense corporate competition caused coal company profit margins to shrink, which undermined their ability to confront new commercial challenges. Finally, the weakening of British hegemony ushered in a new era of escalating tensions between the era's most powerful states. Rising geopolitical rivalries then prompted a shift toward oil-powered ships, airplanes, and motor vehicles. This confluence of systemic factors was already initiating a global energy shift toward petroleum, when the forces of war swept across the world and greatly accelerated this transition.

A similar convergence of dynamics brought the international oil system into crisis later in the century. Social tensions, this time manifested as labor and nationalist movements, emerged in many oil-exporting nations. Meanwhile, the entry of new companies into the oil system undermined oligopolistic order and set the stage for a major wave of expropriations. The geopolitical framework that had contained these forms of social and commercial unrest also began to weaken. Setbacks in Vietnam undermined U.S. hegemony and opened the way for aggressive political challenges to western interests. Under the combined impact of these social, commercial, and political pressures, the international oil system entered its first period of serious crisis.

To identify the social roots of unrest in the petroleum system, it is necessary to begin with an analysis of the distortions that tend to accompany oil-based forms of development. While petroleum sectors often generate impressive levels of financial wealth, it is now widely recognized that this wealth is a decidedly mixed blessing.[1] In country after country, serious structural imbalances have been created by oil bonanzas. Rising oil revenues have often fostered overreliance on imported goods, for instance, which has caused local industry and agriculture to atrophy. With anemic growth in nonoil sectors, the revenue generated by petroleum has grown increasingly crucial to social and economic welfare in many exporting nations. But this wealth has often been monopolized by foreign companies, or by local elites who utilize privileged access to oil rents to sustain insular regimes. These developmental characteristics help explain why authoritarianism and unrest are so often fostered in oil-exporting countries, even while a certain kind of freedom is fostered in core nations that consume the resource.

Given these kinds of volatile developmental trajectories, it is no surprise that many oil-producing nations have witnessed periodic social upheavals. With oil workers caught in the middle of such conflictive environments, it is also no surprise that they have frequently become central participants in struggles to change the conditions under which oil is extracted and marketed. The fact that they have been employed in key revenue-generating industries has also often given them a uniquely powerful position from which to exert economic and political demands.[2]

As early as 1900, demonstrations against Standard Oil's properties in Romania brought about improvements in working conditions. Strikes against foreign companies in Peru, Trinidad, and Bolivia in subsequent years won important concessions as well. Widespread labor unrest was a key factor leading to the successful nationalization of the Mexican oil industry in 1938, while rising oil worker militancy in Venezuela during the same period strengthened that nation's demands for concessionary revisions. Middle Eastern oil workers similarly played crucial roles in supporting the nationalizations of the Anglo-Iranian Company and the Suez Canal in the 1950s.[3]

This rising trend of labor and nationalist unrest was given added momentum by the process of decolonization that swept through the global South after the Second World War. Although some oil-exporting countries in Latin America and the Middle East had attained independence prior to the war, the collapse of European colonial empires in the decades after the conflict fundamentally transformed the context within which the international oil system operated.[4] The consequences of this decolonization process reverberated with particular intensity through international oil sectors in the 1960s and 1970s.

Political independence lent added momentum to the nationalist and anti-imperialist movements that had been building in Latin America and the Middle East. A system that specialized in the transfer of inexpensive oil resources from peripheral to core nations, and that was dominated by western multinational corporations and their home states, became increasingly anachronistic in a new era of political liberation. A generation of leaders in the periphery rode to power on the crest of broad-based popular movements that demanded greater control over important domestic resources like oil.[5] The result was a convergence of labor and nationalist mobilizations that came to threaten the stability of the international oil system.

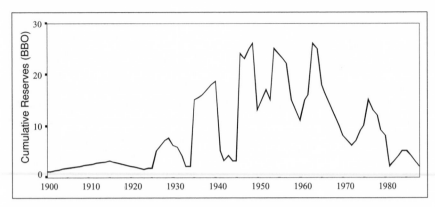

FIGURE 6.1. Discoveries of Oil Reserves, 1900–1988. 5-Year Moving Averages
Sources: Campbell (1991); Campbell & Laherre (1998).

Political independence, combined with rising labor and nationalist unrest, brought the western-dominated international oil system under increasing strain in the late 1960s. At the same time, new competitive dynamics began to undermine the commercial order that had been maintained by
the major oil companies for decades. Discoveries of prolific new reserves
of oil began to slow, which heightened demand for access to the massive
fields of the Middle East. These reserves therefore attracted the attention
of an increasingly large number of new oil companies in the 1960s, and
acute forms of competition that had been staved off in previous decades
began to spread through the region. By the end of the 1960s, the corporate
governance structure that had regulated the international oil system had
been fundamentally weakened by these new commercial dynamics.

During the early phase of growth in the international oil system, there
were almost continual discoveries of massive new oil reserves around the
world. The widening scope of production opportunities provided by these
discoveries gave multinational oil companies the capacity to relocate capital to regions that offered the most favorable investment climates. Labor
and nationalist challenges could be countered during this period by international capital's capacity to relocate. However, from the mid-1960s onward the number of discoveries of giant oil reserves began to decline (see
Figure 6.1).[6] Given this resource context, new companies from Western
Europe, Japan, and the United States began to engage in fierce competition
with the established majors to get access to these prolific oil fields.

The entry of new oil companies onto the international scene was partly
the result of state-sponsored efforts to develop independent access to oil.
As consumers in Italy, France, and Japan became more reliant on oil, for

instance, state companies were charged with developing independent access to the resource in the Middle East and elsewhere. At the same time, private entrepreneurs also emerged to take on the established majors. J. Paul Getty, for instance, signed an agreement with the King of Saudi Arabia that included much higher royalty payments than anything the majors had ever offered. Overall, it is estimated that around three hundred and fifty oil companies undertook new foreign production or marketing ventures between 1953 and 1972.[7]

One consequence of this proliferation of competing firms was a steady reduction in profit margins in the international oil system. The oil majors had maintained annual profit rates of over 20% through the mid-1950s, but by the late 1960s investments in foreign oil were yielding average returns of around 10%, comparable to profits of most manufacturing companies but lower than what the oil majors had been accustomed to.[8] In response, the major oil companies attempted to reduce their payments to producer states. A disconnect therefore emerged, in which the established majors were becoming more severe while newcomers were striking more generous deals with host nations. Hostility toward the majors therefore intensified in the Middle East and Latin America during the late 1960s.

Host governments eventually began to take concerted action to resist increased pressure from the major oil multinationals. In 1960 representatives from the Middle East and Latin America had created the Organization of Petroleum-Exporting Countries (OPEC) to defend their common interests. By the late 1960s this organization was playing a crucial role in publicizing details about the more lucrative contracts being offered by new oil companies and in coordinating resistance to the price cuts that the majors were attempting to impose on host countries.[9]

In the late 1960s social and commercial pressures were clearly mounting in the international oil system. It was precisely at this time that the U.S.-backed geopolitical framework, which had contained radical unrest in the international oil system, was fundamentally weakened. The event that first brought U.S. power into serious crisis came in January 1968, during the Tet offensive in Vietnam. The ability of North Vietnamese forces to launch surprise attacks all across South Vietnam and seize tactical advantage from the United States called into question the ability of the United States to militarily impose its will on peripheral nations. Setbacks in Vietnam also caused the costs of the conflict to escalate, which helped erode U.S. dominance over world financial affairs. Also, the brutality of American attacks in Southeast Asia generated increasingly fierce opposition, thereby undermining the ideological position of the United States at home and abroad.[10] Major setbacks in the Vietnam War therefore played a central role in the erosion of U.S. hegemony and the geopolitical order that sustained the international oil system.

The most immediate effect of the Vietnam War on the petroleum system came with an important shift in U.S. military strategy. Battered by setbacks in Southeast Asia, the Nixon Administration declared in 1969 that henceforth the United States would rely primarily on indigenous military forces to protect American interests in the developing world. The United States increased its sales of military hardware to Israel, Iran, and Saudi Arabia, but refused to station a permanent military contingent in the region.[11] This shift to providing arm's-length assistance to client regimes weakened the security framework that had contained radical threats to the international oil system since the end of the First World War.

The financial difficulties faced by the United States as a result of the Vietnam War also had direct economic repercussions on the international oil system. With an economic recession assailing the U.S. economy and rising competitive challenges coming from Western Europe and Japan, the Nixon administration shifted to a more aggressive international energy policy. Whereas the United States had long worked to ensure that affordable oil resources flowed to its key allies, by the late 1960s the declining hegemon was demanding more compensation for favorable allocations of world oil reserves (Bromley 1991). This shift heightened the need for European and Japanese state oil companies to gain their own access to oil, which intensified competitive dynamics and further undermined commercial stability in the oil system.

Faced with rising social unrest, intensifying commercial competition, and a weakened geopolitical framework, the western-dominated system that had dominated oil production and distribution since the turn of the century became vulnerable to crises generated by conjunctional events. The first major threat to stability came with the Six Day War of 1967, in which the Israeli army engaged with Egyptian, Syrian, and Jordanian military forces. Although Israel quickly defeated its adversaries, the conflict had lasting consequences for the region's oil industries. The Suez Canal was made impassable for a number of years, and petroleum shipments were disrupted. More importantly, resentment against Israel and its allies in the United States and Western Europe intensified in the Arabian world.[12]

In this already volatile context, the 1973 Arab-Israeli War proved to be the catalyst for a fundamental transformation of the international oil system. The conflict led to immediate embargoes of sales of Arabian oil to the United States and the Netherlands. OPEC also orchestrated a 400 percent increase in world prices for oil between October 1973 and June 1974, in part to punish western governments for the assistance they provided to

Israel. Also, a wave of nationalizations spread through the international oil system.

Prior to 1973 governments in countries such as Mexico, Iran, Libya, and Algeria had seized control of foreign-owned oil properties. While these expropriations were seen as threats to the multinationals, they remained isolated events within an otherwise stable international industry. However, after 1973 a wave of expropriations swept through the Middle East and Latin America that overwhelmed foreign companies and their allied states. Indeed, by the late 1970s over 75 percent of international oil properties had been nationalized.[13] Whereas earlier nationalist regimes could be isolated or overthrown, there was little that could be done by already weakened companies and western states to prevent this coordinated expropriation of oil reserves.

By the mid-1970s the international oil system had been radically transformed by an intersection of social, commercial, and political crises. The system that had regulated oil sectors in the Middle East and Latin America for decades was overturned, and peripheral countries took formal control over the production and pricing of their resources. The multinationals continued to operate in most of these countries, but only on the basis of limited contracts that carefully specified services to be carried out on behalf of the host governments. Oil prices also became much more volatile in this new era, while shipments of the resource were shown to be vulnerable to military conflict and political unrest.[14] For the first time, consumers in core countries were forced to recognize that they had become dependent on an energy resource that originated from nations that were escaping the control of western corporations and their home governments. As crises rippled through the international oil system, governments and citizens in the global North struggled to devise responses to these new realities that had seized hold of the global energy system.

Systemic Dynamics and the Initiation of a New Energy Shift

Just as had happened earlier in the international coal system, increasing volatility in the oil system prompted important changes in energy sectors across the world. Governments throughout the core took the lead in attempting to confront these challenges by increasing subsidies to conventional energy industries, supporting research on new energy systems, and implementing conservation measures. Private entrepreneurs took advantage of tax incentives to improve energy efficiencies and increase investments in a variety of energy technologies. And, in what was perhaps the most important change of all, the oil shock energized a variety of new environmental organizations throughout the global North. For the first time, groups of citizens in Western Europe, North America, and Japan became directly engaged in debates over which energy systems should be supported and which should be abandoned. Although the impact of these political, commercial, and social dynamics ended up having only a temporary effect on the global energy system, they nevertheless demonstrated again that significant changes in energy trajectories can be achieved in a short period of time.

In examining the responses to the first oil shock, it is important to begin by noting that, for many analysts of the day, the shock portended a major shift in global political and economic relations. The OPEC intervention was taken by many as evidence that the unequal relationships of power and prosperity that existed between core and peripheral zones of the world-system could be fundamentally transformed.[15] In the flush of success that followed the embargo and price adjustments, OPEC ministers stated that they would support efforts to create new cartels in other commodities sectors and that they would push for improved terms of trade for all developing country products. Moreover, as OPEC members began to receive increased revenue for their oil, they redirected some of these funds to neighboring countries that were suddenly faced with much higher oil import bills (Hallwood and Sinclair 1982). These actions by OPEC, combined with rising calls for a new international economic order and the final defeat of the United States in Vietnam, seemed to herald a new era of rising power in the global South.

In the face of what seemed to be a coordinated challenge headed by OPEC, the initial response from core nations was remarkably disorganized and incoherent. A division emerged in Western Europe between countries like the Netherlands, which tried to shut OPEC oil out, and France, West Germany, and the United Kingdom, all of whom tried to strike new deals with Arabian exporters to get oil flowing to their consumers again. Japan, for its part, signed new deals with almost every country that was capable of exporting oil. The United States, meanwhile, had virtually no coherent response at all in 1974, since the country was embroiled in the Watergate scandal.[16]

Within the next few years, however, core states developed something of a common response to the new crisis in the international oil system. Improved coordination was partly achieved through the creation of the International Energy Agency (IEA), which brought core states together for the first time in a formal way to devise joint responses to energy challenges.[17]

First, these core states attempted to develop new sources of oil that were outside the OPEC orbit. This led to a significant increase in public and private investments in petroleum projects in the North Sea, the Gulf of Mexico, Alaska, and West Africa. Next, they sharply increased investments in domestic energy industries. Nuclear power plants received massive new funding in the United States, France, and Japan, for instance, while natural gas projects expanded in the United Kingdom, the Netherlands, and the United States. New subsidies were also directed to coal mines in the United Kingdom, the United States, France, and Germany.[18] By the end of the 1970s the natural gas and nuclear power industries had been set on particularly rapid growth trajectories as a result of these combined public and private investments.

In addition to investing in established energy industries, many core states also increased their expenditures on energy-related research and development (R&D) projects. In fact, R&D expenditures in IEA countries grew from a very low level in 1970 to over $10 billion a year from 1977 through the early 1980s.[19] It is important to note that most of this money went to nuclear and fossil fuel sectors. In fact, over 61 percent of the $75 billion spent by IEA governments on energy R&D between 1974 and 1980 went to nuclear power projects; around 13 percent of this money went to fossil fuel projects; and another 12 percent went into storage and transmission research. Only 8 percent of this sum was directed to renewable energy systems, and only 5 percent went to conservation efforts (see Figure 7.2 on page 151 for an historical overview of trends in R&D expenditures).

Even though renewable and conservation projects received only a small share of all government R&D funding, this still represented a significant expansion in state support. Over the period 1974–80, IEA countries spent a total of $10 billion on alternative energy projects. The U.S. government led the world in these investments, providing over 60 percent of the total funding in renewable and conservation expenditures for the period. Western European countries jointly provided 27 percent of global funding for these R&D projects, while Japan followed with 13 percent of the total.

These patterns in state R&D expenditures clearly reveal the continuing influence wielded by fossil fuel and nuclear power companies in core countries during the 1970s. However, the R&D data also demonstrate that support was growing for more sustainable responses to energy dilemmas. New social pressures, this time exerted by environmental movements, were clearly beginning to have an effect on the trajectory of global energy industries by the 1970s.

Environmental concerns about the impacts of energy production and consumption have a long history. Attempts to control smog generated by the burning of coal, for instance, were undertaken as early as the thirteenth century in London. By the late nineteenth century, increasingly severe smogs were engulfing cities in Europe and North America. Particularly severe disasters occurred in the Rhine Valley, Belgium, Pennsylvania, New York, and London in the postwar period. Finally, after a series of major public health catastrophes, governments took steps to address urban air pollution generated by fossil fuel industries. For instance, the smog event that killed over four thousand people in London in 1952 spurred passage of the world's first national-level clean air legislation in Britain in 1957. Similar efforts to reduce urban smog were made in other European and North American cities during the 1960s and 1970s.[20]

During these decades, scientific evidence began to mount that pollution from fossil fuel consumption was impacting rural areas as well. For instance, emissions of sulphur dioxide and nitrogen oxides from coal-fired electric utility plants were shown to cause acid rain, which harmed lakes, rivers, and vegetation across broad regions. Ecosystems in England, Germany, the United States, and Japan began to exhibit widespread signs of damage as a result of acid rain. In response to this scientific evidence, national regulations designed to control air pollution were enacted in Western Europe, North America, and Japan in the mid- to late 1960s. As a result, moderate reductions in urban and regional pollution rates were achieved during the subsequent decades in core countries.

There were many intersecting social dynamics that led public officials to enact these kinds of environmental regulations. Scientific groups throughout the world, for example, played a key role in raising public awareness about the damage caused by commercial energy industries. Private corporations occasionally adapted more sustainable practices as a result of regulatory pressure and economic calculations regarding the benefits of efficiency. Meanwhile, the relative affluence of the postwar period allowed middle-class citizens to turn their attention to health and quality-of-life concerns, which put pressure on government officials to address pollution problems.[21] Along with these moderate environmental approaches, there also arose a variety of movements that used grassroots tactics to protest ecological destruction in general, and conventional energy industries in particular.

The post–Second World War period saw a steady rise in citizen mobilizations against large-scale energy projects. As early as 1956, for instance, residents of Detroit picketed against a nuclear power plant that was being constructed near their homes. In the mid-1960s activists regularly protested against dam projects in the American West, oil spills in the North Sea, and coal pollution in the Ruhr. The late 1960s and early

1970s then saw a qualitative jump in the number of people participating in antipollution and antinuclear campaigns. For instance, by the mid-1970s at least 300,000 people in Britain had joined some organization that dealt with conservation or antipollution issues, while over 20 million people in the United States had become members of environmental organizations by that time. And millions of people around the world participated in activities associated with the first Earth Day, which was held in April 1970.[22]

These citizen-based forms of environmentalism, when combined with mainstream campaigns, helped encourage governments to include support for renewable energy projects and conservation efforts in their array of responses to the first oil shock. In North America, Western Europe, Japan, and some other countries, governments increased investments in R&D, offered enhanced subsidies, and created tax incentives designed to stimulate alternative energy sectors. Oil and utility companies also became important partners in many alternative energy enterprises as part of their efforts to diversify investments in a new era of risk. Energy activists also began to champion specific alternative energy systems in many communities across the world.

Although investment patterns differed from country to country, the largest share of renewable investments flowed into long-established hydroelectric, biomass, and geothermal projects. There had already been a boom in dam construction underway across the world when the first oil shock came. Even more capital was invested in hydroelectric programs after 1973, especially in the developing world, in efforts to cope with escalating oil-import expenses.[23] There was also a push to increase the use of biomass resources. Brazil, for instance, increased the production of sugar cane alcohol for use in its transportation sector, while the United States expanded its corn-based ethanol program. Efforts were also made to make better use of biomass resources generated by agricultural, forestry, and garbage industries.[24] In countries like Italy, Iceland, Hungary, Japan, and the United States, investments in geothermal energy systems also grew significantly after 1973. By the end of the 1970s large-scale geothermal facilities were generating heat and electricity in these countries at prices that were almost competitive with conventional industries.[25]

In addition to these investments in established renewables, efforts were also undertaken to expand the market penetration of newer, more innovative energy systems. Specifically, public and private investments generated a flurry of research and entrepreneurial activities in wind, solar, and fuel cell sectors in the 1970s. For instance, the newly created Department of Energy and the National Renewable Energy Laboratory sponsored research that significantly improved the efficiency of American solar and wind systems. Across the Pacific, Japan's Project Sunshine funneled money into solar thermal and photovoltaic arrays. In Denmark, the United Kingdom, Argentina, and Australia, state-sponsored projects generated expansions in the commercial use of wind power after 1973.[26]

Solar energy systems in particular flourished after the first oil shock. Solar thermal and photovoltaic technologies each have surprisingly long histories of experimentation and commercial use. Solar thermal devices that harness the sun to heat water to high temperatures date from Swiss experiments in the 1760s. Interestingly, a steam engine powered by solar-heated water was demonstrated at the 1867 Great Exhibition in Paris. However, the true commercial introduction of solar thermal systems took place in the early 1900s in the United States. Between 1909 and 1918 an American entrepreneur installed over four thousand solar water heaters in homes across the southern United States.[27] Photovoltaic cells that generate electrical power from sunlight, meanwhile, were first designed in 1839 by a French physicist. A few photovoltaic arrays were tested by private companies in the 1950s, while by the 1960s they were being introduced into virtually every U.S. and Soviet space vehicle and satellite.[28]

After the oil price hike, commercial sales of solar thermal and photovoltaic systems rose sharply in the United States and Japan. Over the decade of the 1970s, around one million solar water heaters were sold in the United States while at least 700,000 were sold in Japan.[29] Meanwhile, worldwide production of photovoltaic-generated electricity rose quickly, coming to exceed 21.3 megawatts of installed capacity in 1983. In that year, U.S. companies controlled 60 percent of the photovoltaic market, while the Japanese share was 23 percent and European companies produced most of the rest.[30]

This rapid expansion in solar systems was driven not only by government and corporate sponsorship, but also by the fact that, across the industrialized world, hundreds of thousands of people became convinced that they could increase their energy independence and protect the earth by installing solar systems in their homes. Environmentalists, survivalists, antinuclear activists, and a whole host of other kinds of people built and installed their own solar systems. Indeed, by the mid-1970s a full-fledged solar advocacy movement had sprung up in the United States and Japan.[31] In the eyes of many of these alternative energy activists, a relatively rapid shift away from fossil fuels and nuclear power and toward a solar-powered economy could be kick-started by grassroots efforts and modest government support.

The final energy system that received increased attention during the 1970s, and that would spur its own grassroots advocacy movement in future decades, was the hydrogen-powered fuel cell. Fuel cells resemble common batteries in that they rely on chemical reactions to produce electricity. In their most environmentally pristine form, fuel cells are injected with hydrogen and oxygen, which, when exposed to a catalyst, react to generate electricity. In hydrogen-powered fuel cells, pure water is the only byproduct generated by this electrochemical reaction. Recent engineering advances have also permitted the use of fuels such as methanol, natural gas, and oil in fuel cell systems—and these fuel cells do generate some carbon dioxide emissions. But because fuel cells can attain much higher efficiency rates than conventional engines, their emissions levels are greatly reduced even when they are powered with chemicals derived from fossil fuels. Because of their environmental benefits, it has often been claimed that hydrogen-powered fuel cells can become a key energy technology in a more efficient, environmentally sustainable energy system.[32]

The first demonstration of a fuel cell was conducted in 1839 by the British scientist William Grove. Almost a century passed before a practical cell was developed by Francis Bacon, an engineer at Cambridge University. Bacon's general design was then adopted by American companies like General Electric and Pratt & Whitney, which provided hydrogen-powered fuel cells to the U.S. space program from the late 1950s onward.[33] Years of engineering experience in state-sponsored space programs significantly improved the reliability of hydrogen-powered fuel cells, so that by the time of the first oil shock the technology was relatively mature. During the 1970s joint ventures between government agencies and private companies explored how the fuel cell could be used in terrestrial applications in the United States, Japan, and Western Europe.[34]

In an associated development, efforts to scale up the commercial production of hydrogen for use as a fuel were also undertaken. Hydrogen had already been used to power vehicles in the 1920s and 1930s in Europe. Zeppelins relied extensively on hydrogen, while somewhere between one and four thousand trucks were converted to the fuel during the Second World War in Germany in an effort to make up for shortages of oil. Interest in using hydrogen as a fuel died with the influx of cheap oil, but the first oil shock revived hydrogen R&D projects. The United States increased support for hydrogen research in the mid-1970s, while West Germany and Japan initiated research programs a few years later. By the late 1970s the International Energy Agency was also coordinating a multicountry hydrogen research program.[35]

In addition to all these efforts to expand old and new energy systems, there were also important steps undertaken to reign in energy consumption in many countries. Across the industrialized world, governments imposed

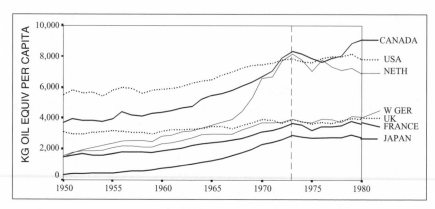

FIGURE 6.2. Per Capita Energy Consumption in Selected Countries, 1950–1980
Sources: See Appendix A.

new codes on construction industries that led to improved efficiencies in commercial and residential buildings. Companies accelerated their shifts away from energy-intensive manufacturing and toward service-oriented activities. Voluntary agreements induced companies to boost the efficiency of major household and business appliances, and rebates encouraged consumers to buy these new appliances. Also, in the United States at least, new speed limits were enacted that reduced oil consumption in the transportation sector. Finally, informational campaigns encouraged people to turn off lights, turn down thermostats, and take other steps to reduce their personal energy consumption.[36]

These efforts to achieve structural and behavioral changes in energy consumption did have important effects in the years after the first oil shock. For one, the exponential growth rate that had taken hold of the international oil system after the Second World War was halted. Meanwhile, an analysis of per capita energy consumption ratios shows that, in some countries, notable improvements in efficiency were achieved in a short amount of time. As Figure 6.2 shows, the Netherlands and Canada saw significant reductions in energy consumption ratios, while Japan and France achieved moderate reductions. Even in the United States, the United Kingdom, and West Germany, per capita consumption rates were held steady through the 1970s. Reducing the global rate of oil consumption and reigning in energy use in affluent nations represent significant achievements—ones that again demonstrate that global energy systems do not follow set trajectories but are instead responsive to human interventions.

The decade that followed the first oil shock was clearly a time of tremendous innovation and diversification in energy industries across the world. Under the combined effect of increased state support, corporate investments, and grassroots enthusiasm, conventional and unconventional energy systems not tied to OPEC oil underwent important growth spurts. For the first time, some government officials, industrialists, and citizens in affluent nations began to look for ways to reign in energy consumption. While the successes of these energy conservation efforts proved to be temporary in most countries, they still represent a historic departure from the dominant ethos that had reigned throughout the history of the modern global energy system. At long last, demands for perpetual expansions in energy industries were countered, however hesitantly, by a call from some quarters for limits to energy consumption.

At the close of the 1970s, the global energy system appeared to be in a new period of transition. The oil embargoes and price hikes orchestrated by OPEC spurred states and corporations to intensify their efforts to diversify their sources of energy. While nuclear and fossil fuel industries received the bulk of state and corporate attention, there was also an expansion in state support for alternative energy systems. Meanwhile, the consolidation of a multifaceted environmental movement in the global North brought new groups of citizens into the debates over which kinds of energy systems should be expanded and which should be abandoned. Although the transition that began as a result of these multiple pressures would ultimately falter, the reforms achieved in the 1970s continue to stand as a testament to the ability of social forces to alter the trajectory of the global energy system in a relatively short time.

A World Historical Interpretation of the Second Oil Shock

Oil embargoes and price hikes were not the last of the crises to assail the international oil system in the 1970s. In fact, at the end of the decade an even more radical event spread turmoil through the global energy system. Specifically, the Iranian revolution of 1979 brought the downfall of a prowestern regime and the consolidation of a new government that was openly hostile to the United States and its allies. The turbulence accompanying this revolution disrupted petroleum shipments and generated severe oil price shocks, each of which had negative repercussions on world-economic stability. Because of these events, it was revealed that the world remained susceptible to the destabilizing impact of unrest in the Middle East. Coming after a decade of sustained efforts to reduce reliance on this region's oil supplies, this was a particularly ominous discovery for those concerned with the long-term stability of the global energy system.

As with the first oil crisis, this second period of turmoil can be seen as having arisen out of a convergence of destabilizing social, commercial, and political factors. An analysis of this particular case of regime change reveals vulnerabilities that are shared by many other oil-exporting countries. The Iranian revolution therefore represents a type of crisis that is likely to again assail the international oil system in the coming decades.

The roots of the Iranian Revolution stretch back to the period before the First World War, when oil was first discovered in what was then known as Persia. The region had already experienced regular incursions by European military forces during the nineteenth century. The opening of the first Persian oil well in 1908 and the onset of war heightened foreign interest in this strategically located region. During the First World War, wide sections of the country were devastated as Turkish, German, Russian, and British troops engaged in repeated clashes. At the end of the war, the British sent troops to occupy oil fields and other key areas. After a long period of failed attempts to create client regimes, the British backed a military strongman—Reza Shah—who was able to bring an end to rebellions in the region.[37] In their interventions, the British therefore helped create two enduring features of modern Iranian society that would have long-lasting repercussions: they established western control over the country's most important resource, and they fostered the creation of an authoritarian political system that was tied in important ways to foreign powers.

From 1921 until his abdication in 1941, Reza Shah worked to construct a centralized state that could contain social unrest while programs intended to modernize Iran's military, industry, and agriculture were pushed forward. Although the Shah clashed with the British over the terms of oil concessions in the 1930s, he remained dependent on the United Kingdom for assistance until his regime collapsed under the weight of the Second World War. During this conflict, Iran saw a repeat of what had happened in the first war, only with a shift in the key occupying force. After confrontations took place between German, Soviet, and British forces, the United States intervened with tens of thousands of troops. Key oil fields were seized, and U.S. diplomats and financial advisors set about trying to create a stable postwar government that would be allied to western interests. They turned to Reza Shah's son, Mohammad Pahlavi, who was appointed to the throne in 1941 at the age of 21. Though widely viewed as an ineffectual leader, Shah Pahlavi was willing to support the terms of western oil concessions. Furthermore, it was thought that, if given enough western support, the new Shah could continue pushing Iran along the path of centralized modernization his father had begun.[38] Once again, western control over Iran's oil and the fortification of an authoritarian regime tied to foreign interests seemed securely in place.

As has already been discussed, the Shah's pro-Western orientation was briefly interrupted by the nationalization program promoted by Mohammad Mossadegh in 1951. Although Mossadegh's nationalization was strongly supported by the people of Iran, the United States and Britain joined forces to sponsor a coup that drove Mossadegh from office and reversed the expropriation. By 1953 Shah Pahlavi had been put back in charge of the government, oil properties had been returned to foreign ownership, and a more intense phase of authoritarian rule began in Iran. Relying on massive western financial and military aid, the Shah's security services set about repressing sources of political and social opposition. Iranian labor organizations, which had spearheaded over two hundred strikes in support of the nationalizations, were particularly hard hit, while anti-Western politicians and intellectuals were also silenced.[39] The continuities in postcolonial Iranian society—western ownership of oil and political authoritarianism—were therefore reinstated by the combined intervention of the United States and the United Kingdom.

In the two decades that followed, the Shah's government worked to strengthen its security apparatus. The intelligence service SAVAK was created in 1957 with the aid of the Central Intelligence Agency and Israel's Mossad, and the Shah also purchased advanced weaponry for his growing army. The British pullout from the region following the Suez debacle, a perceived buildup of Soviet power, and rising internal unrest all combined to prompt a massive increase in U.S. military and financial aid to Iran. Even after the Shah supported OPEC embargoes and price hikes, the Nixon Administration continued to bolster the regime in the face of what seemed to be more radical dangers. By the early 1970s the Shah's government—with the full backing of the United States—had come to rely on kidnappings, torture, and executions as regular parts of its response to internal dissent.[40]

In addition to the resentments generated by an increasingly repressive government, distortions generated by oil-based development began to produce social tensions within Iran. Steady revenues from the petroleum sector had financed expansions in public education, improvements in sanitation services, and some job creation in the country's urban centers during the 1960s. However, the massive influx of money that followed the price hikes of 1973 had more problematic effects. Rising inflation eroded the incomes of ordinary Iranians, while the central government engaged in increasingly ostentatious expenditures. The Shah's fascination with military weaponry led him to purchase fleets of fighter jets, helicopters, tanks, and other advanced systems. Meanwhile, a growing share of development money was shifted away from projects designed to help regular citizens and toward elite-oriented health clinics, shopping centers, and resort communities. By the mid-1970s, Iranian society had become extremely polarized, with a small, wealthy elite engaging in western-style forms of conspicuous consumption and a burgeoning number of rural and urban citizens struggling in the face of declining economic prospects.[41]

As in many other oil-rich countries, the revenues flowing to the Shah's regime fostered social polarization, economic distortions, and political insularity. With rural agriculture in decline, migration to urban areas accelerating, and inequalities intensifying, the context was ripe for opponents to mobilize against a ruling elite that was widely perceived to be corrupt, autocratic, and subservient to the West. Given the repressive tendencies of the Shah's government, leaders of dissident groups were excluded from gaining peaceful access to the political system. As a result, increasingly radical movements began to coalesce within three important social groups that experienced declining fortunes under the Shah. The first group to raise a serious challenge to the established order was Shi'ite clerics and their followers. They were then joined by merchants from the bazaar economy. And, at the very end of a long process of growing tensions, oil workers carried out a wave of strikes that brought the Shah's regime into terminal crisis.[42]

The most tenacious resistance to the Shah's government came from networks of Shi'ite clerics that had long dominated rural communities and that greatly strengthened their presence in urban shanty towns in the postwar period. The Shi'ite tradition within Islam is characterized by an emphasis on worldly asceticism, egalitarianism, and strict adherence to the strictures of the Koran. In the twentieth century, a growing cultural divide emerged between regular Iranian citizens, over 90 percent of whom are Shi'ite, and an increasingly westernized ruling elite, which adopted fundamentally different moral orientations. In the face of the rising social inequalities and cultural dislocations wrought by modernization programs, a growing number of impoverished citizens across Iran began to place their allegiance in clerics who articulated fundamentalist versions of the Shi'ite world view, and who criticized the authoritarian, western-oriented elite that had come to control the country.[43]

The cleric that gradually emerged as the leader of this oppositional movement, Ayatollah Khomeini, was relatively unique in his consistent criticism of the Shah's regime. Khomeini's public pronouncements against the government began in 1962 in response to a land reform program that reduced the property holdings owned by Shi'ite religious organizations. The next year, the Ayatollah condemned the Shah for rigging parliamentary elections and for being too subservient to the United States (Abrahamian 1980). In the massive popular uprisings that followed, the Shah's security forces killed over one thousand demonstrators and forced Khomeini into exile in Iraq.

Khomeini's exile did not stop the Ayatollah from criticizing the Shah. On the contrary, it seemed to have facilitated this campaign. Clerics who remained in Iran were subject to the direct repression of SAVAK, and so they tended to mute their public denunciations of the government. Khomeini, by contrast, was relatively protected across the border in Najaf, Iraq, and so he was able to give full vent to attacks on what he perceived to be a corrupt, dictatorial, and atheistic regime. These declarations were then distributed by networks of Islamic activists on tape cassettes, so that by the 1970s Khomeini had become the most influential Ayatollah in Iran. His demands for greater social justice and religious piety, combined with his attacks on the Shah and his western allies, had particular resonance among impoverished Islamic students, urban residents, and lower-class workers who had few opportunities for upward mobility in the inflation-ridden economy that had been generated by petroleum-based development.[44]

A worsening economic situation in the mid-1970s finally forced the Shah to impose a structural adjustment program designed by the International Monetary Fund, which ended up alienating another key group in Iran. The price controls and import restrictions that were included as part of the economic austerity program had a direct, negative impact on the fortunes of the merchants that controlled most of Iran's retail trade. For generations these merchants had maintained independent guilds that regulated prices within the world of the bazaars. As part of what he claimed was a campaign against profiteering, the Shah tried to dissolve these traditional guilds and turn control of the bazaars over to wealthy business allies. This provoked strong resistance within the merchant class and solidified their alliance with clerics who were by now openly calling for the overthrow of the regime.[45]

By 1978 key preconditions for the onset of a revolutionary crisis had coalesced in Iran.[46] A broad-based, multiclass movement had emerged to challenge the legitimacy of a personalistic, authoritarian regime. At the same time, international circumstances constrained the ability of the regime's main foreign sponsor, the United States, to protect the Shah's government. Post-Vietnam opposition to foreign interventions ruled out a large-scale deployment of American military personnel. A bolstered presence of Soviet forces in the region, and in Afghanistan in particular, also

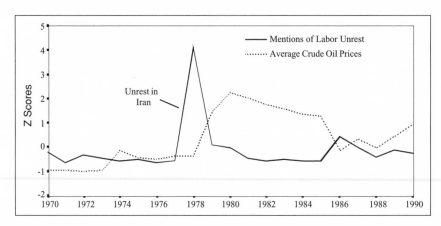

FIGURE 6.3. Labor Unrest and Price Shocks in the International Petroleum System
Sources: Price series: IEA (1998); Labor unrest series: See Appendix C.

limited the U.S. option of responding with direct military support for the Shah. At the same time, prerevolutionary crises were also mounting in Nicaragua and El Salvador, and so covert operations resources were stretched thin. The result was that, though the Carter Administration sent some additional military hardware and financial assistance to Iran, the Shah was left to face rising domestic threats largely on his own.

The Shah's regime responded to the evolving crises in an increasingly incoherent fashion, shifting abruptly from the use of severe repression to the granting of concessions. For instance, early 1978 brought a massacre of religious students and large-scale arrests of clerics and merchants. By the fall, though, many of these clerics and merchants were suddenly released from jail (Parsa 1988). Regardless of which tact the regime took, Ayatollah Khomeini's popularity grew and larger social mobilizations against the government occurred.

In September 1978 the final key social group—the oil workers—joined the mass protests against the Shah. The petroleum workers went out on strike, thereby immobilizing operations at key oil refineries, petrochemical facilities, and fuel distribution centers. These lengthy strikes not only disrupted domestic oil sales, but they also halted Iranian oil exports and had a broader impact on the international oil system. As Figure 6.3 shows, there is a strong association between rising labor unrest in Iran—as measured by the World Labor Research Group Database described in Chapter 5—and global oil price hikes.[47] This is a particularly clear demonstration that social unrest in a key peripheral oil-exporting country can have a significant, direct impact on global energy markets.

In response to these damaging strikes, the Shah's regime continued to follow an erratic course of action. Modest wage concessions were offered, but when these were rejected by the oil workers the government sent the army to arrest strikers and occupy refineries. However, oil workers barricaded themselves inside many facilities, while sympathy strikes spread through other industries. In fact, by January 1979 it was estimated that over three million industrial workers were on strike against the Iranian government. Meanwhile, peaceful demonstrations of more than a million people were held in Tehran and in other major cities.[48]

In mid-January 1979 the Shah left the country on what was intended to be a temporary trip. The new premier who was left in charge of the domestic situation immediately pledged to free all political prisoners, reign in the security forces, and ease economic austerity measures. However, on the first of February Ayatollah Khomeini returned to the country and named his own prime minister. A few clashes then ensued between military units loyal to the Shah and armed insurgents, but power was ultimately transferred to the revolutionary forces in a remarkably rapid and peaceful manner.

The new regime quickly received assurances from the Soviet Union that support would be given in the case of any effort by the United States to foster a coup. Although the new government at first seemed interested in maintaining at least neutral diplomatic relations with the United States, events took a more radical turn when Islamic militants captured the U.S. embassy in Tehran and seized fifty-two American hostages.[49] This crisis, combined with a failed attempt by U.S. special operations forces to rescue the captives, played a central role in ensuring that Jimmy Carter was not reelected. Radical social unrest in Iran therefore had a direct impact on the political trajectory of the United States, which, as we will see, then had long-term consequences for the evolution of the global energy system.

In its first incarnation, the revolutionary government in Iran was made up of a broad coalition of Islamic and non-Islamic groups. Direct elections for Parliament and for the presidency were instituted, and independent political parties were allowed to form. However, rising popular support for Khomeini—combined with postrevolutionary crises—set the stage for the consolidation of power in fundamentalist Shi'ite hands. The intensification of tensions between Iran and the United States played a role in this process, while the onset of the Iran-Iraq war then accelerated the Islamicization of the government. In September 1980 Iranian militants shelled Iraqi towns along their mutual border. Soon afterward, Saddam Hussein unleashed a full-fledged invasion that was meant to quickly overwhelm the new Iranian regime. However, popular loyalty to Khomeini turned out to be deeper than outside analysts had anticipated. Though revolutionary Iranian forces had little experience using modern military systems, their use of human wave tactics against better-armed Iraqis managed to halt the invasion, even at a tremendous cost in lives. Far from defeating their enemy, the Iraqi attack accelerated the consolidation of power in the hands of Khomeini and his fundamentalist allies (Keddie 2003). By the mid-1980s the Islamic government had proved that it could withstand attacks from external enemies and that it would be an enduring political force in the Middle East.

While the OPEC-driven oil disruptions of the early 1970s represented one kind of challenge to the stability of the international oil system, the Iranian Revolution was the manifestation of a much more radical and destabilizing form of unrest. The ownership and price adjustments that occurred in the early part of the decade were certainly painful for western interests, but they could ultimately be accommodated within the existing capitalist system. However, the seizure of an oil-rich country by Islamic clerics, who came to articulate an increasingly severe anti-Western world-view, was a threat of an altogether different order of magnitude. For the first time in history, a peripheral country with major oil reserves had been taken over by a regime that pledged to use its newly found wealth to mount direct challenges to the United States and its allies.

In addition to the broad geopolitical implications of the Iranian Revolution, there were important social and commercial ramifications as well. The collapse of the Shah's regime demonstrated the internal vulnerabilities of heavily populated, oil-exporting nations. From 1979 on, the stability of countries like Iraq, Saudi Arabia, Nigeria, and Venezuela could no longer be taken for granted. Moreover, the power of strategically located oil workers to generate domestic and international economic turmoil was revealed in very clear terms. In sum, the international oil system in particular, and the global energy system more generally, emerged from the 1970s in a very uncertain state.

The Stalled Energy Shift

With the emergence of a new, more radical kind of threat in the Middle East, it might have been expected that support for a shift to a more diversified, sustainable energy system would have grown in core countries. After all, rising labor and nationalist unrest in oil sectors had already set in motion the earliest stages of a shift to a more diversified and sustainable energy system in the mid-1970s. Should not renewed unrest in the late 1970s have provided added impetus to this shift?

The reaction to the second oil shock followed a different pattern, however. One central response by advanced industrial states was to try to expand reliance on nuclear power. Because a huge share of government and private resources was invested in nuclear programs, financing for more environmentally sustainable projects withered. Moreover, grassroots support for alternative energy systems had to be diverted to efforts to prevent the wholesale expansion of nuclear power. This state-driven effort to expand nuclear power therefore played a key role in stalling the shift to a more diverse, sustainable energy system in the 1980s. Leading core states and companies also focused on expanding fossil fuel industries, and so even more resources were diverted away from environmentally sustainable energy technologies. Given the combination of a failed effort to shift toward nuclear power and intensified support for fossil fuels, the global energy system emerged from the 1980s with a strong reliance on coal, oil, and natural gas, but with few other alternatives under serious development.

The steadfast support for nuclear power and fossil fuels that dominated energy policy in the 1980s went hand in hand with a larger effort by the United States and Great Britain to reassert control in a world that seemed to have grown increasingly chaotic. U.S. hegemonic influence was severely battered by the defeat in Vietnam, by OPEC's interventions, by revolutions in Iran and Nicaragua, and by the Soviet invasion of Afghanistan. Economic recession was assailing many advanced capitalist economies, while many peripheral countries were enjoying improved prices for the commodities they were exporting. In response to these challenges, the new Reagan and Thatcher governments orchestrated a counterrevolution designed to reinstate U.S./U.K. political and economic dominance over the Soviet Union, on the one hand, and the global South on the other.

A central component of the Reagan/Thatcher counterrevolution was the intensification of Cold War rivalries with the Soviet Union. This competition was partly manifested as a renewed nuclear arms race and partly as a stronger willingness to intervene militarily in nations throughout the global South. These military-political strategies each had direct implications for energy policies pursued by the United States and the United Kingdom during the 1980s.

Soon after taking office, Reagan announced that he would be deploying cruise missiles to Western Europe to counter threats posed by Soviet SS-20 missiles. Reagan and Thatcher also initiated massive programs to expand their conventional and unconventional weapons arsenals. As part of these armament programs, a new wave of state funding surged into research on advanced atomic systems. The military-industrial complexes that had been born in the United States and Britain during the Second

World War, and that had pushed for expansions in civilian nuclear power sectors in the 1950s and 1960s, were therefore reinvigorated in the early 1980s. With revived support for nuclear systems within the Reagan and Thatcher administrations, the stage was also set for ambitious programs to expand civilian nuclear power industries in the United States and Britain. Meanwhile, Japan and France continued doggedly along in their efforts to reduce dependence on foreign oil by expanding their own nuclear power sectors.[50]

We can again look at state expenditures on energy-related R&D activities to see the relative commitment level offered to various energy systems by core countries. Once again, this analysis shows very clearly that the advanced industrial nations that were part of the IEA[51] continued to give preferential support to nuclear power. Of the $123 billion dollars spent by IEA countries on energy-related R&D activities during the period 1980–90, fully 60 percent ($74 billion dollars) went into civilian nuclear power research. Fifteen percent of the R&D money went to fossil fuels, while 7 and 4 percent went to renewable and conservation activities, respectively (IEA 2004).

In the aggregate, the patterns of R&D expenditures between the periods 1974–80 and 1980–90 show quite a bit of constancy. The proportion of money spent each year on nuclear power remained a huge proportion of government R&D support. Meanwhile, fossil fuel expenditures remained largely steady while funding for renewables and conservation R&D went down somewhat. Within these aggregate figures, though, were distinct national shifts. The Reagan and Thatcher administrations, for instance, slashed funding for alternative energy systems and conservation efforts while they significantly increased nuclear and fossil fuel research funding.[52] The opposite pattern occurred in Scandinavian countries and West Germany, where state support for nuclear research declined and funding for renewables increased. In France and Japan, state support for nuclear sectors remained steady throughout the whole 1974–90 period (see Figure 7.2 on page 151 for a broader overview of expenditure patterns).

The fact that state support for nuclear power remained relatively strong through the 1980s is a testament to the impermeability of many government energy agencies to public sentiment. We have already seen that grassroots mobilizations against nuclear power stations had begun in the 1960s and 1970s in the United States and Western Europe. These public campaigns were intensified in the wake of the 1979 Three Mile Island accident and after the 1986 Chernobyl disaster. Across Western Europe, massive public demonstrations occurred throughout the 1980s against the potential deployment of nuclear missiles and against civilian nuclear power stations.[53] The creation of many Green parties in Europe was stimu-lated by these antinuclear campaigns, and by the late 1980s environmental or Green parties had won significant seats in the parliaments of West Germany, the Netherlands, Belgium, Sweden, and Finland (Rudig 1991). It is no accident that these countries began to reduce support for nu-clear sectors in the mid-1980s and increase funding on renewable energy systems.

Although large demonstrations against nuclear power also took place in the United States, Britain, France, and Japan, federal governments in these countries continued to champion the technology through the 1980s. This is partly because the political systems of these countries operate in ways that keep minority parties excluded, and partly because powerful public/private interest groups had been created that helped sustain support in the face of growing public opposition. There is ample evidence showing that pub-lic opinion had swung against nuclear power in these countries in the 1980s. For instance, surveys conducted in the United States revealed that about twice as many people opposed nuclear power as supported it throughout the decade. Strong public concern also emerged in Japan after 1986, and by 1990 a survey revealed that 90 percent of the public felt uneasy about nuclear power. Meanwhile, surveys carried out in Britain after 1986 revealed that strong majorities opposed expansions in nuclear power.[54]

It was not only grassroots activism and public opinion that weighed in against nuclear power. Energy economists in the United States and Great Britain also began to express concern about the commercial viability of the technology. One American analysis calculated that, if only construc-tion, operating, and maintenance costs were taken into account, the U.S. civilian nuclear power industry had received about $492 billion in private and public investments during the period 1950–90. Given this conser-vative estimate of invested costs (decommissioning, waste handling, and other costs were not included), electricity from nuclear plants had ended up costing U.S. consumers an average of 9 cents per kilowatt hour, or more than twice the price of electricity produced in conventional plants over that period. An analysis of the British nuclear industry conducted in the late 1980s revealed a similar picture; massive public and private in-vestments had yielded electricity that was two to three times more expen-sive than fossil fuel–generated power.[55] In both countries, these extremely high costs—combined with growing estimates of decommissioning and waste handling charges—kept private utility companies from assuming full responsibility for nuclear power plants as privatization plans moved forward.

Under the combined pressure of grassroots activism, general public opposition, and commercial uncertainty, the massive expansion in nuclear power that had been planned in many countries was held in check. This important achievement reveals a couple of important features of the contemporary global energy system. First, we again see that social mobilizations can play a critical role in influencing the trajectory of large-scale energy industries. Second, the example of nuclear power reveals the danger of excessive state interventionism in energy planning. It was only through the exertion of centralized federal power that civilian nuclear power industries were initiated at all. Once underway, the political and corporate interest groups tied to nuclear projects were hard to stop, even as the safety, environmental, and commercial limitations of the technology began to become apparent. While virtually all civilian nuclear power industries have been contained, this has only come after a massive diversion of public and private resources away from more promising technological options. Clearly, future decisions about which energy systems to prioritize should emerge from healthier negotiations between political, corporate, and public interest groups.

If one central component of the Reagan/Thatcher counterrevolution involved a reintensification of the Cold War—with its associated support for nuclear power—another key part of the counterrevolution focused on efforts to control dissident regimes in the global South. These efforts at hegemonic reconstruction were driven partly by the need for the Reagan and Thatcher administrations to appear strong to domestic constituencies. They were also responding to threats to commercial interests in certain regions. Of most relevance to this analysis, these more aggressive military and financial interventions were also motivated by the desire to establish a stronger degree of influence over the international oil system.

The most dramatic manifestation of the Reagan/Thatcher counterrevolution against the global South came in a series of overt and covert military operations carried out in Latin America and the Middle East in the 1980s. Interventions in places like the Falklands and Grenada allowed each new administration to project an image of strength to domestic and international observers. American interventions in Central America, meanwhile, were justified using anticommunist arguments, though an absence of evidence of Soviet support for revolutionary movements in the region suggests that U.S. involvement was driven instead by a desire to protect commercial interests in the region. Whatever their motivations, though, interventions in Latin America were not driven to any large extent by energy resource concerns. The same cannot be said for U.S. and British campaigns in the Middle East. In this region, overt and covert military engagements were driven more fundamentally by a desire to reestablish privileged access to oil and gas.

The clearest example of oil-motivated intervention in the Middle East came with the growing rapprochement between the Reagan Administration and Iraq that occurred in the early 1980s. For over a decade, U.S.-Iraq relations had been strained by Saddam Hussein's reliance on military and financial aid from the Soviet Union. However, when the Iranian Revolution transformed a key ally into a hostile enemy, the United States began to fortify its relations with Hussein. Declassified documents reveal that by 1982 the Reagan Administration had begun providing logistical and material support to Iraq in its war against Iran. In visits to Hussein undertaken in 1983 and 1984, Donald Rumsfeld provided private assurances that the Reagan Administration supported Iraq even as that country was being condemned for using chemical weapons against Iranians.[56] Over the next five years, the United States provided military equipment and satellite images to Iraq's military, and U.S. forces even directly attacked Iranian ships and oil installations. In addition to this support for Hussein, massive increases in weapons sales to Saudi Arabia, Kuwait, and other countries in the region were approved by Reagan in an effort to resolidify U.S. influence in the region.

In addition to these military interventions, the Reagan/Thatcher counterrevolution against the global South had an important financial dimension as well. It has already been noted that the price of oil rose significantly in the 1970s. These price rises generated a massive pool of capital that seeped through OPEC countries and then back into the global banking system. As financial institutions struggled to find investment outlets for this money, lending terms became increasingly flexible. Countries with domestic oil reserves that could be used as collateral were courted by international lenders, and even elites in many oil-poor nations were permitted to borrow heavily in this cash-rich environment. In most cases, oil-poor countries had to dedicate a large part of this borrowed money to finance mushrooming energy import bills.[57] Given this confluence of factors, a large number of developing countries became heavily indebted in the 1970s.

Although these high debt rates appeared prudent in a context of strong global liquidity, a change in the international financial climate generated a crisis of mammoth proportions for developing countries. This change originated with the monetarist counterrevolution that the Reagan and Thatcher administrations orchestrated in the early 1980s.[58] Whereas the United States and the United Kingdom had followed relatively conservative fiscal policies in the previous decade, the new administrations allowed debt levels to mushroom as they simultaneously cut domestic taxes and engaged in massive military build-ups. The result was that global surplus capital was increasingly sucked into the United States and Britain, leaving developing economies struggling in a much harsher financial environment. Unable to service their debt repayment obligations in this new climate, country after country in Latin America, Africa, and Asia descended into debt crises in the early 1980s, from which many have yet to emerge.

The debt crisis was a key mechanism that allowed core countries to reimpose control over a periphery that had achieved important degrees of independence in the 1970s. By the 1980s the IMF and the World Bank were imposing structural adjustment and liberalization programs that once again reduced developing countries' control over their domestic resources. These programs had an important impact on international oil and natural gas industries. If the 1970s had seen a wave of nationalizations in oil sectors, by the 1980s all but the wealthiest oil exporters were forced to open their petroleum and gas sectors to foreign corporations.[59] By the mid-1980s, oil and gas exports were increasing from indebted countries like Angola, Argentina, Cameroon, Gabon, Indonesia, Malaysia, and Mexico. With assistance from multilateral organizations like the IMF, multinational energy corporations were able to accelerate the process of geographic diversification that had begun after the first oil shock.

This diversification in oil and gas sources played a key role in undermining the power of OPEC. While in 1974 OPEC's share of world oil production had been at 38 percent, by 1985 the cartel controlled only around 20 percent. Moreover, OPEC countries still needed to sell their oil resources to affluent countries. By the early 1980s, core countries were able to use their power as consumers to sideline oil from particularly radical countries—like Libya and Iran—and reward more moderate countries with larger purchase contracts. This induced countries like Nigeria to cheat on their OPEC quotas, which then caused markets for OPEC stalwarts to shrink. For a number of years Saudi Arabian oil ministers attempted to hold OPEC together in the face of these production and market shifts. But finally, in 1986, Saudi Arabia opened its wells to full capacity in order to recapture market share for its own oil. World oil prices quickly fell back to price levels that were, in real terms, similar to what had existed just prior to the first oil shock.[60]

In sum, the counterrevolution led by the Reagan and Thatcher administrations set off a chain reaction that ultimately caused a rebalancing of political and economic power in the international oil system. Although OPEC countries retained control over their own domestic production decisions, they were forced to contend with a more competitive oil system that had been created in large part by the debt crisis. Moreover, with politically aggressive governments in power in the United States and Britain, corporations were willing to undertake new oil and gas investments in many regions around the world that had been judged too unstable in earlier decades.

The combination of burgeoning oil and gas production in new areas, weakened OPEC influence, and falling world oil prices set the stage for a reversion back to heavy reliance on imported oil and gas resources in most core nations. Whereas energy consumption rates had been held constant or decreased in many advanced industrial countries in the 1970s, the 1980s saw countries like the United States, Great Britain, Canada, and Australia abandon conservation efforts (Meyers and Schipper 1992). Simultaneously, government support for alternative energy technologies was cut back sharply in these countries. Although funding for solar and wind systems remained steady in Japan and Continental Europe, multinational efforts to commercialize hydrogen-based fuel cells also declined.

At the end of the 1980s most industrial countries had returned to energy consumption patterns that were reminiscent of the early 1970s. Reliance on imported oil and gas resources had again intensified, while few viable alternatives to fossil fuels were under development. The massive investments made in nuclear power failed to generate a viable energy system based on this technology due to widespread social, commercial, and technical problems. Unfortunately, huge infusions of public and private capital in nuclear power sucked money away from other energy development projects, and so a transition to a more diversified and sustainable energy system stalled. Instead, the global energy system would move into an increasingly precarious state as the twentieth century drew to a close and a new century began.

7 Toward a Sustainable Energy System

A s the world begins the twenty-first century, the long-term viability of the global energy system is increasingly being called into question. This era's most important energy industry, based on oil, is being repeatedly wracked by major security crises. Indeed, the cycle of terrorist attacks and counterattacks that has gripped Afghanistan and Iraq—and is echoed in Israel and the Occupied Territories—brings a new level of danger to the world's most prolific oil-producing regions. Moreover, revelations that nuclear technologies and materials are proliferating through state and nonstate channels reveal the inherent dangers of this energy system, especially in an era of growing security threats. Finally, rising temperatures across the world demonstrate quite clearly that global warming is a reality, and that a concerted response is urgently required.

A new global energy shift, away from destabilizing energy systems, is clearly necessary if major geopolitical, commercial, social, and environmental difficulties are to be avoided in the coming years. This chapter explores the possibilities for carrying out this kind of shift by situating possible future trajectories in their broader world historical context. It begins by examining the threats that are posed for the global energy system, and the opportunities that are opened, as the United States continues its hegemonic decline. Once contemporary energy crises have been described, the chapter then turns to an analysis of the new energy systems that are capable of providing sustainable solutions to these dilemmas. The study then concludes by discussing how world historical dynamics can help foster a new shift toward a more sustainable and equitable global energy system.

Declining U.S. Hegemony and Contemporary Energy Crises

The last decade of the twentieth century ushered in what appeared to many to be an era of revitalized U.S. hegemony and global prosperity. The collapse of socialist governments in Eastern Europe and the rapid defeat of Iraq in the first Gulf War established the United States as the world's preeminent military power. The United States also emerged as a leader in the development of the era's new high-tech industries, while it was also able to help foster stability in key energy-producing countries. Finally, the nation developed a multilateral approach to global politics and finance that enhanced the ideological legitimacy of the United States in key states around the world. In many respects, therefore, the United States seemed to be well on its way to constructing a new hegemonic order that would last long into the twenty-first century.

Even in the midst of this apparent renaissance, though, many world-systems researchers argued that the United States remained trapped in a trajectory of hegemonic decline. While all agreed that the United States had historically unparalleled levels of military power at its disposal, scholars like Wallerstein (1992), Arrighi and Silver (1999), and Boswell and Chase-Dunn (2000) steadfastly maintained that this power was ephemeral. They each predicted that processes of weapons proliferation, global economic restructuring, and social polarization were undermining the capacity of the United States to create lasting order on a global level.

These arguments seem particularly prescient today, following the terrorist attacks and unilateral armed responses that have occurred in the early twenty-first century. While the United States retains an important military edge, it has become overextended due to its precipitous invasions of Afghanistan and Iraq. Moreover, its ideological claims to be pursuing global objectives of freedom and democracy ring hollow in many communities across the Middle East, Europe, Asia, and elsewhere following its preemptive war against Iraq.

The unraveling of U.S. hegemony is linked in many important ways to emerging crises in the global energy system. There are certainly nonenergy factors in play, of course; however, a series of challenges arising out of established energy industries is contributing to global disorder. A review of these crises suggests that the global energy system must be significantly reformed if severe catastrophes are to be avoided in the coming decades.

The most immediate energy-driven threats emanate from the maturing international oil system. It is important to note that five Persian Gulf countries—Iran, Iraq, Kuwait, Saudi Arabia, and the United Arab Emirates—contain around two thirds of all proven reserves of crude oil. Even though oil sectors in much of the rest of the world have been opened to multinational corporations in recent decades, the boom of production taking place in these regions will probably peak within twenty years. A shift back toward reliance on oil controlled by key Persian Gulf producers—and the Organization of Petroleum-Exporting Countries

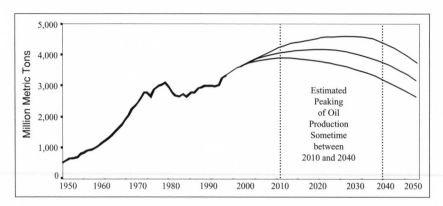

FIGURE 7.1. Petroleum Production Forecasts Based on Varying Estimates of Recoverable World Oil Reserves
Sources: Production Data: Appendix A; Projectections: Campbell (1991); USGS (2000).

(OPEC) more generally—will inevitably take place. Indeed, according to forecasts from the U.S. Energy Information Administration, almost half the world's oil will come from OPEC nations by 2025.[1]

Though large reserves of oil remain in certain regions of the world, there is a growing concern about the approaching peak of global oil production. As Hubbert (1962) first argued, the production of nonrenewable resources generally follows a bell-shaped curve of rising and then declining output. If one can estimate the total volume of a resource and combine that with expected consumption trends, it is possible to forecast when production will peak with some degree of accuracy.[2] Once this peak occurs, all future extraction takes place at a declining rate and higher cost.

When a Hubbert-style analysis is applied to contemporary global oil reserves, the results are less than encouraging. There is disagreement within the community of petroleum geologists about how much oil remains to be extracted using techniques that are currently in use or likely to be deployed in the coming years. Some pessimistic analyses claim that there are less than two trillion barrels of conventional oil in the world, while the most optimistic studies (provided by the U.S. Geological Survey) claim that around three trillion barrels remain in the ground.[3] Given these varied estimates of recoverable oil and assuming a typical consumption growth rate of 2 percent per year, it is estimated that conventional oil production will peak somewhere between 2010 and 2040 (see Figure 7.1). Even increasing the resource estimate to an implausible five trillion barrels would only postpone the peak to around the year 2070. These estimates of the medium-term prospects for oil suggest that the world-economy will become steadily more reliant on oil reserves held by a few countries, and that these reserves will grow more expensive. For the first time in modern history, the world is likely to be facing real material constraints on the availability of the era's key energy resource in the relatively near future.

In addition to the rising costs of oil and the strengthened market position for OPEC that will flow from this material reality, a large number of oil-importing nations are being exposed to new sources of oil-driven instability. Though the United States and its allies retain conventional military primacy in key zones of oil production, it is becoming clear that their ability to contain security crises in places like the Middle East is diminishing. This is partly a lingering consequence of earlier efforts by the United States to fortify its influence in the region. Throughout the 1980s and 1990s, the United States and its allies sold huge volumes of conventional weaponry to client states in the Middle East and Central Asia in efforts to fortify western influence.[4] Also, recall that during the 1980s the United States turned a blind eye to Iraq's use of chemical weapons against Iran. More recently, state control over nuclear materials has weakened as a result of Pakistani proliferation activities and continuing deterioration in nuclear facilities across the former Soviet Union. As a result, the regions with the richest reserves of oil are now the sites of the some of the most perilous concentrations of conventional and unconventional weapons in the world.

The fact that political and social conflicts continue to escalate in this hypermilitarized context heightens the likelihood that the international oil system will be disrupted in the coming years. The Israeli/Palestinian conflict continues unabated, while anger at the United States and its allies has deepened in the Middle East following the Iraq war.[5] At the same time, Saudi Arabia, Kuwait, and the United Arab Emirate are exhibiting many of the same characteristics of regime insularity, financial corruption, and social polarization that emerged in Iran in the 1970s (Okruhlik 1999).

The fact that these regimes have allied themselves to varying degrees with the United States lends added fuel to the campaigns of groups like al Qaeda, which are attempting to destabilize regimes in the region and the industries they rely on. Given the prevalence of powerfully destructive weapons, radical groups are gaining an unprecedented capacity to wreak havoc on energy industries in the Middle East and elsewhere.

These regional problems are likely to be compounded by broader geopolitical tensions in the coming decades. Burgeoning oil consumption in China, for instance, is already drawing that nation strongly into Central Asian and Middle Eastern oil markets.[6] If oil remains the world's preeminent source of energy in the coming decades, nations of the global South will come into direct competition with nations of the global North for access to the resource. The material inequalities embedded in the global energy system could therefore become a more potent source of conflict than those that existed before the Second World War.

Given these problematic trends in the international oil system, many countries have increased their efforts to expand natural gas industries. Indeed, the decade of the 1990s has witnessed a particularly rapid growth in the global production of natural gas. As will be discussed below, this expansion in natural gas production is a development that is to be welcomed. Since this is the least polluting of the fossil fuels, greater reliance on natural gas could allow the world to avoid shifting wholesale toward the highly polluting and dangerous resources of coal, unconventional oil, and nuclear power.

Unfortunately, increases in the production of these more problematic sources of energy are also being registered. Coal output has undergone renewed growth in China, the United States, Germany, India, Australia, and South Africa. At the same time, efforts are underway in Canada, Venezuela, the United States, and China to extract energy from heavy oil, shales, and tar sands. Global production of these unconventional sources of oil is projected to grow steadily in the coming decades.

There is also a renewed determination in some countries to undertake yet another expansion of nuclear industries. Proposals have recently been submitted to regulatory agencies in the United States to permit the opening of a new reactor around 2008. Meanwhile, electricity production at existing nuclear facilities in the United States, Japan, and France has steadily increased. New government investments are also being made in nuclear fusion projects, in the hope that this elusive energy technology may one day approach commercial viability. Meanwhile, North Korea and Iran have recently insisted that they will expand their civilian nuclear power sectors.[7]

A shift toward heavy reliance on coal, unconventional oil, and nuclear power may lie in our future. Unfortunately, extensive scientific evidence demonstrates that this outcome would be quite hazardous to the well-being of human society and the world's ecological systems.

One of the most serious dangers facing contemporary world society is the pollution generated by conventional fossil fuel and nuclear industries. It has been estimated that more than one billion people throughout the world live in cities where air pollution levels exceed healthy levels. The human consequences of this energy-generated pollution can be quite significant. In the United States, for instance, it is thought that at least 28 percent of the urban population is exposed to harmful levels of exhaust particulates, which is a level of exposure that is thought to cause the premature death of around forty thousand Americans each year. In the developing world, conditions are even more extreme. In Mexico City, high levels of pollution are estimated to cause over 6,500 deaths each year. Meanwhile, over fifty-two thousand people in thirty-six Indian cities are thought to have been killed by air pollution in 1995 alone. In China, air pollution is estimated to cause anywhere from 170,000 to 280,000 deaths each year. On top of the human toll registered in these figures, there are growing financial costs as well. In developed countries air pollution is estimated to cost around 2 percent of gross domestic product (GDP); in developing nations such pollution can cost anywhere from 5 to 20 percent of GDP. On a global scale, the health costs of urban air pollution are thought to approach $100 billion annually.[8]

Rising pollution from fossil fuel industries is also threatening the stability of the world's climate. According to a study prepared by hundreds of scientists working for the Intergovernmental Panel on Climate Change (IPCC 2001), the greenhouse gas emissions already in the world's atmosphere will bring about a rise in global temperatures of between two and ten degrees Fahrenheit by the end of the century. This rise in temperature will produce increasingly severe weather events, including droughts, floods, and massive storms, in various parts of the world. Melting polar ice will also raise sea levels and potentially disrupt ocean currents that moderate temperatures in Europe and elsewhere.

Although some of these effects may take a century to manifest themselves, there is growing evidence that certain climatic shifts can occur in the space of only a decade. A study by the National Academy of Sciences (2002) reported that abrupt climate changes have occurred in the past, and that another abrupt change may now be underway. Consider the fact that the five hottest years since 1861 have occurred in 1995, 1998, 2001, 2002, and 2003. During these years, massive forest fires have swept through temperate and tropical areas, while hundreds of thousands of people have died in unprecedented heat waves. In the summer of 2003, for instance, nearly fifteen thousand people died in France during one extended period of heat and humidity.[9] While climate change is already having harmful impacts on people and ecosystems, a report commissioned by the U.S. Department of Defense outlines even more ominous scenarios that may await us. As Schwartz and Randall (2003) describe, it is conceivable that within a decade shifts in the world's climate could dramatically decrease food production, disrupt fresh water cycles, and contribute to social and political conflict in many regions of the world.

While the evidence of danger continues to mount, there is still no comprehensive effort under way to limit global emissions of greenhouse gases. In March 2004 the European Union committed itself to abiding by the

Kyoto accords, but countries like the United States still refuse to ratify the treaty. Instead, the United States continues to lead the world in per capita emissions of energy-related greenhouse gases. At the same time, developing nations such as China and India are also generating larger emissions. As a result, global emissions have risen throughout the 1990s, and now far surpass the target levels established at the 1997 Kyoto summit.[10]

In the face of this accumulating evidence of danger, coal and oil industry advocates argue that new pollution-reduction, carbon-sequestration, and greenhouse gas offset practices can allow for an expansion in coal and oil consumption without the creation of major environmental problems.[11] However, scrubbers installed on fossil fuel plants generate huge quantities of toxic sludge that are difficult and costly to handle. Underground carbon containment will also be quite expensive and is unlikely to be able to store large quantities of gas for long periods of time. Finally, recent analyses of forest growth call into question the idea that carbon emissions in one area can be offset by reforestation activities in another. A key problem is that young trees do not appear to absorb as much carbon as mature trees, and so new growth may not provide the absorptive capacity that had originally been envisioned.[12] Far from becoming carbon sinks, the world's remaining forests instead show signs of becoming increasingly vulnerable to catastrophic fires as global climate change accelerates.

Advocates of nuclear power have also seized on concerns about global warming to argue that their industry offers a viable solution to greenhouse gas problems. However, the multiple environmental and security threats posed by this technology appear to rule out a worldwide shift toward this energy system. As has already been discussed, countries that build civilian nuclear power industries also have opportunities to develop the expertise needed to deploy offensive nuclear weapons. Unless the world wants to become more fully nuclearized, the proliferation of civilian nuclear sectors to new countries must be halted. Serious problems have also arisen in countries that already have the technology. No nation has yet devised a reliable and affordable system for containing the waste produced by civilian reactors. Instead, the massive waste facilities that dot the world have become enticing targets for terrorist attacks. It has also become clear that management and security systems meant to ensure the containment of fissile materials can collapse during periods of political and economic turmoil. This has been demonstrated in the former Soviet Union, where at least eighty pounds of weapons-grade uranium and plutonium have been stolen during the last decade.[13]

Though most of this nuclear material has been retrieved, it is thought that at least some has ended up in the hands of illicit arms dealers. Recent warnings from U.S. and European security agencies indicate that al Qaeda operatives are attempting to acquire radioactive materials that can be used in dirty bombs. These threats have become so real that, in early 2004, the U.S. government sent scientists into the streets of a handful of major American cities to monitor for radiation.[14] No matter how strong the advocacy work is in support of expanding civilian nuclear sectors, one detonation of a dirty bomb or an attack on a nuclear facility will almost certainly turn world opinion irreversibly against this energy system.

If the international oil system is entering into a sustained period of vulnerability and if coal, unconventional oil, and nuclear power industries are hazardous alternatives, what options are there for satisfying global energy demands? There are a variety of energy systems that have the potential to reduce greenhouse gas emissions, increase efficiencies in conventional energy sectors, and deliver relatively clean power to communities around the world. Many of the threats posed by current energy trajectories could be eased by shifting more rapidly toward natural gas, unconventional methane deposits, and renewable energy systems. At the same time, concerted efforts to reduce energy demand, especially in nations that consume far more than the world average, would also reduce tensions generated by the global energy system.

While the policies of the United States do not currently favor a rapid deployment of more sustainable energy technologies, there may be unexpectedly positive consequences that emerge from the current crisis. As the historical record shows, shifts toward new energy systems happen most readily during periods of turmoil. If the most severe kinds of disasters can be averted, then systemic dynamics should again favor far-reaching transformations in global energy industries. Indeed, as the conclusion of this study will argue, a further weakening of U.S. hegemony is likely to be associated with rising crises in oil and nuclear power industries. This is likely to stimulate efforts by governments, corporations, and communities across the world to deploy a cluster of more sustainable energy systems that have been under development for decades, and that are now ready for mass production.

Technological Solutions to Contemporary Energy Crises

Just as the last decade of the twentieth century brought a period of peace and prosperity, it also brought hope that major environmental problems might at last be addressed on a global level. By the late 1980s, scientific and popular awareness of growing threats to the world's ecosystems and climate began to generate pressure for an international response to environmental problems. The result was the 1992 Earth Summit, where over one hundred heads of state gathered to sign agreements to protect biological diversity and the earth's climate. Indeed, by signing the Framework Convention on Climate Change, the world's political leadership committed itself in a nonbinding way to reduce many of the greenhouse gases being generated by the global energy system.

In addition to this promising international development, there were signs that new leaders in many affluent nations were set to push forward on more aggressive conservation and renewable energy projects. The election of the Clinton/Gore administration in the United States, for instance, heightened expectations that the world's most prolific consumer of energy would enact domestic reforms and support the roll-out of a host of renewable energy systems. Similarly, the ruling party of Japan announced major new energy conservation and development projects in the early 1990s, while a variety of Western European governments also committed themselves to expanding their sustainable energy sectors. A number of major corporations, including oil companies, also began proclaiming their support for solar and wind systems. Finally, a new upsurge in grassroots movements in support of alternative energy systems began to occur in countries throughout the world.[15]

As described below, advances were made in bringing a host of new energy systems to the verge of commercial viability in this favorable context. However, the achievements of the 1990s were far from revolutionary. After a proposal for a modest tax on energy failed in the United States and the Senate voted 95–0 to reject a climate change treaty, the Clinton Administration abandoned further efforts to significantly change American energy policy. Whereas support for new energy systems grew in Japan, increases in conventional and nuclear energy consumption grew even faster. Meanwhile, state support for renewables failed to grow appreciably in Western Europe in the 1990s. Energy efficiencies therefore either held steady or declined in most countries, while greenhouse gas emissions from the advanced industrial world continued to increase. The United States was a particular culprit in this regard, as increased sales of SUVs and burgeoning electricity consumption led to a steady growth in greenhouse emissions.[16]

One way to evaluate national governments' commitment to energy conservation and renewables is to turn again to an examination of the research and development (R&D) expenditures made by public agencies. Figure 7.2 provides an overview of public R&D expenditures by member states of the International Energy Agency (IEA) for the period 1974–98 (IEA 2003).

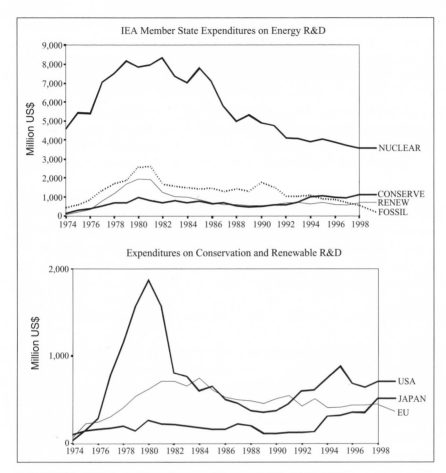

FIGURE 7.2. IEA State Expenditures of Energy R&D, 1947–1998.
Source: International Energy Agency (2004).

The top figure shows the patterns of state investments in all energy systems, while the bottom figure focuses on conservation and renewable expenditures in key countries. As the figures reveal, there has been a slow decline in state R&D support for nuclear power since the mid-1980s, though this technology still receives more than 50 percent of all public R&D support on an annual basis. There has also been a modest decline in fossil fuel R&D funding. Indeed, in the mid-1990s fossil fuel support finally dropped below that of R&D spending on conservation technologies. Meanwhile, spending on conservation efforts has shown modest growth in the 1990s. On the other hand, aggregate public R&D expenditures on renewable energy systems remained virtually flat throughout the decade.

Turning to an analysis of spending on R&D for conservation and renewable systems in particular countries, we see that the U.S. government has almost doubled its annual outlays on these activities since the early 1990s. Though government support for these types of R&D remains significantly below levels attained in the late 1970s, the United States has kept a healthy lead in these efforts since the early 1990s. Japanese state support for conservation and renewable systems, meanwhile, has more than quadrupled since 1990, while the combined expenditures of the countries that now make up the European Union have at least held steady since the mid-1990s.

There was clearly a modest shift in energy-related R&D expenditures in the 1990s, with slight gains in support for conservation and renewables balanced in part by reductions for nuclear and fossil fuel systems. Still, what is most remarkable is the overwhelming commitment to nuclear power, and the failure of governments to significantly shift their energy-related R&D expenditures after the 1992 Earth Summit. From a historical perspective, however, this lack of change in state support for energy industries is not surprising. As we have seen, major shifts in state support for energy industries have occurred most readily during eras of acute crisis, and the 1990s were certainly not such a period. Instead, the late twentieth century resembles the late nineteenth century, with the peace and prosperity of each mature period of hegemony fostering strong reliance on that era's key energy resource.

Remember, though, that while the late nineteenth century was witnessing the full blossoming of the international coal system, new energy systems were also beginning to take shape. By the time severe turmoil assailed the coal system, a new energy regime based on oil was well positioned to enter into rapid expansion. Commercial developments had also advanced far enough with internal combustion and electrical systems that these technologies were able to undergo massive growth during the turbulent middle decades of the twentieth century. As we turn to a review of the status of various energy industries at the turn of this century, the historical parallels are clear. Just as oil was amassing the capacity to supplant coal as the world's key resource one hundred years ago, natural gas and methane resources are now emerging as strong medium-term alternatives to overreliance on oil. Similarly, the fuel cell and a cluster of renewable energy technologies are on the verge of mass production, so they too are well poised to undergo rapid expansion during the coming decades.

Of all the alternatives to carbon-intensive and hazardous resources, natural gas is the most viable candidate to provide large volumes of relatively clean energy in the coming decades. Natural gas is the least environmentally harmful of all hydrocarbons, with a given unit of energy from gas generating about 40 percent less carbon dioxide than oil and 75 percent less than coal. This lower level of emissions has already led to the introduction of natural gas vehicles in cities like Los Angeles, Santiago, and Shanghai, marking only the beginning of what promises to be a wide-scale diffusion of gas-based technologies throughout the world.[17] By the end of the 1990s, natural gas was poised to surpass coal as the world's second most significant source of commercial energy, and all energy forecasts predict that gas will undergo the most rapid growth of all resources in the coming decades.

The total amount of conventional natural gas thought to be ultimately recoverable lies somewhere in the range of five thousand to six thousand trillion cubic feet. Of these reserves, around 39 percent are located in the Middle East. Another large proportion, about 31percent, is located in Russia, while the rest is distributed quite liberally across the world (U.S. Energy Information Administration 2003). Given anticipated gas consumption rates, the production peak for this resource is quite a bit further away than that for oil. With its cleaner characteristics, its wider geographic distribution, and its longer consumption horizon, conventional natural gas therefore provides a relatively good alternative to coal, unconventional oil, and nuclear power.

There is another advantage to harnessing gas resources for commercial energy production. The primary component of natural gas is methane, which is a powerful greenhouse gas. Uncontrolled releases of methane from leaky pipelines, oil wells, coal mines, sewage systems, and landfills currently contribute significantly to global warming. If these releases of methane can be stopped, and the energy contained in the methane captured for commercial use, then the benefits could be enormous (Hansen et al. 2000). It is even possible that this strategy of harnessing methane can be expanded to incorporate unconventional deposits of the gas. For instance, a number of lakes in tropical zones contain submerged bubbles of methane that can be tapped. Similarly, there are huge volumes of frozen methane in ocean and ice deposits all around the world. While many of these methane hydrates are in polar regions, where extraction is extremely difficult, others are located near industrial centers. Japanese researchers, for instance, are exploring ways of harnessing massive deposits of methane that lie off that country's east coast. The global estimates of methane hydrates are so large that harnessing only about 1 percent of the resource would provide more energy than is contained in all known reserves of conventional natural gas.[18]

There are potential drawbacks to turning more intensively toward unconventional sources of methane, of course. Some undersea deposits could be destabilized when tapped, which could lead to uncontrolled releases of the gas. Surface-level methane extraction can be environmentally destructive as well. In the Powder River Basin in Wyoming, for instance, tens of thousands of wells have been drilled into coal seams to get at methane gas. Not only is the landscape scarred with roads and drill holes, but tremendous volumes of contaminated water have been generated in the methane recovery process.[19] Though extracting embedded methane is a high-impact activity that needs careful regulation, there are low-impact systems that allow methane resources that are already leaking into the environment to be captured. If more of these greenhouse emissions can be harnessed for human use, the transition toward a natural gas energy system could be accelerated while its environmental impact is minimized.

One of the best ways to reduce the impact of natural gas consumption is by integrating fuel cells more fully into this expanding energy system. The electrochemical reactions within fuel cells convert natural gas into useful energy at very efficient rates while generating fewer emissions than traditional combustion systems. Given these advantages, the fuel cell is likely to become the key power device in the natural gas system, just as the internal combustion and steam engines became the key motive technologies in the oil and coal systems.

Developments in fuel cell technologies over the last decade have broadened the horizon for this technology even further. By now, fuel cells have been designed to run on biofuels, petroleum, sewage sludge, hydrogen, and other resources. The environmental advantages and versatility of this technology have generated great excitement, leading to a new wave of rather utopian descriptions of the transformative potential of the device. Just as analysts have at various times predicted that steam engines, nuclear power plants, and solar systems would liberate humanity from toil and poverty, a new generation of enthusiasts has begun to proclaim the revolutionary possibilities of the fuel cell.[20]

There are dangers in overhyping the potential of any energy technology, including the fuel cell. As Romm (2004) points out, excessively optimistic projections of fuel cell sales have already created false expectations about the ease of deploying the technology, which may erode support for the technology among disillusioned policy makers and consumers. And, as with any energy system, there are environmental impacts that must be identified and monitored. For instance, most of the cells require platinum and other relatively scarce metals as catalysts, and so mining in sensitive ecosystems is likely to grow as the fuel cell industry expands. There is also some evidence that hydrogen leaking from fuel cell systems might harm the ozone layer.[21]

Perhaps most importantly, there is the Jevons paradox to contend with. As the British economist William Jevons noted, it is a mistake to assume that more efficient energy systems automatically lead to a decrease in energy consumption. The opposite, in fact, is most often what occurs. More efficient energy transformers lower the per unit cost of captured energy, which then stimulates increased consumption of the resource.[22] It is therefore a mistake to assume that the deployment of fuel cells will save natural gas over the long term. As discussed at the end of this study, true reductions in energy consumption require political and social transformations; they are not caused by energy technologies alone.

What is needed is a balanced evaluation of current developments in fuel cell sectors, so that reasonable projections can be made about the future trajectory of the technology. Fortunately, a series of surveys have recently been conducted on fuel cell markets, which allow us to move away from eulogistic studies of individual companies and toward a more general analysis of the current state of this rapidly evolving industry.

Analysts from the research organization *Fuel Cell Today*, for instance, report that the number of fuel cell systems in operation around the world rose from about 200 in 1990 to over 6,800 in 2003. Approximately 55 percent of these fuel cell systems were located in North America in 2003, while Europe contained about 25 percent and Japan had around 20 percent.[23] At the time of the surveys most of these stationary fuel cells were being tested in laboratories, utility plants, and data centers for their ability to provide back-up electricity. A large number of the systems had also been installed in cars and buses, which were being run as test vehicles by government and corporate organizations. A smaller number of cells had been inserted into computer laptops and other electronic components, where their ability to generate steady, low-level power was being evaluated.

Up through the end of 2003, the vast majority of these fuel cell systems were noncommercial test models. However, a very small number of fuel cells were also being sold on the open market. U.S. companies like Plug Power and GenCore, for instance, were already selling stationary fuel cells designed to replace diesel generators. Meanwhile, a number of industry studies predict that fuel cells are on the verge of mass production and commercialization. Large companies like General Electric, Siemens, United Technologies, and Westinghouse are poised to begin selling stationary cells on a broad scale.[24] A variety of joint partnerships in the automotive world are vying to become the pioneers in commercializing fuel cell cars. Currently the main competitors in this race are the Ballard

Power/DaimlerChrysler/Ford consortium and the General Motors/ Toyota team, though strong efforts are also being mounted by Honda, BMW, and Mitsubishi. A recent automotive industry survey predicted that low-level production of fuel cell cars would begin in 2005, while large-scale sales of the cars would begin in 2008–10.[25] If this timetable is attained, it would mean that the mass production of fuel cell cars will begin almost exactly a century after Henry Ford's roll-out of the Model T set off revolutionary changes in the global auto industry.

One indication that there is a ready market for efficient fuel cell cars is the recent success that has been achieved with hybrid vehicles. With their combination electric/internal combustion engines, hybrids manufactured by Honda and Toyota have generated strong demand among affluent consumers in Japan, Western Europe, and certain regions of the United States. Approximately 150,000 hybrid cars were sold around the world between 1999 and 2003; in fact, the number could have been higher since consumer demand outstripped availability in many markets. A survey conducted in 2003 by J.D. Power & Associates predicted that worldwide sales would accelerate as new hybrid cars and trucks are introduced by companies like Ford, Chevrolet, Dodge, and Toyota. Rising world gasoline prices is one factor that is expected to drive healthy demand for fuel-efficient vehicles in the coming decade.[26] In an echo of what occurred a hundred years ago, when the internal combustion engine emerged as a strong contender to the steam engine, the fuel cell now appears to be on the verge of beginning a competitive race against oil-powered engines.

Just as a host of new electrical systems underwent a spurt of commercialization a century ago, we are now witnessing the dawn of a new era of expansion of modern renewable energy industries. The early twenty-first century is in fact poised to undergo the same kind of transformation that swept through the world in the early twentieth century, when new power lines and electrical generators spread across the world. Only this time, it is possible that the primary energy fed into these electrical systems and heat generators will come increasingly from renewable energy resources.

Just as with the fuel cell, it is important to situate any discussion of modern renewables in its proper context. Remember, for instance, that renewable energy industries (which include hydroelectric, biomass, geothermal, solar, wind, and other modern systems) together provided only around 3.5 percent of the commercial energy consumed in the year 2000. And, as with the fuel cell, almost every modern renewable has suffered from excessively high expectations. Fortunately, recent surveys of modern renewable sectors again allow for a balanced evaluation of the current state and future prospects of these industries.

By far the largest source of renewable energy comes from large-scale hydroelectric dams. These dams provided about 2.7 percent of the world's commercial energy in 2000, while all other modern renewables contributed only about 0.8 percent. According to a survey carried out by the World Commission on Dams (2000), at least forty-five thousand large dams had been constructed by the late 1990s. Though building activities have slowed since their peak in the 1970s, by the end of the century nearly half of the world's rivers had at least one large dam (defined as a structure with a reservoir volume of more than three million cubic meters).

While these dams provide an important source of electricity, especially in the developing world, they have generated many adverse social and environmental impacts. The World Commission estimates that somewhere between forty and eighty million people have been displaced by dams, while many ecosystems and culturally valued sites have been inundated. Maintenance and siltation problems have reduced the power generation capacity of many large dams, while processes of anaerobic decay that occur in many reservoirs is generating high levels of methane gas—and so the greenhouse gas offsets claimed for dams are likely to be overstated. While China and a few other countries continue to forge ahead with ambitious construction projects, most analysts rule out a major increase in large-scale hydroelectric power generation in the coming decades. However, there do seem to be many opportunities for expanding smaller-scale dams in developing countries.[27]

Though growth prospects for hydroelectricity seem uncertain, there is much more optimism about expansion in other renewable sectors. This growth begins from a very small base, of course, since under 1 percent of the world's commercial energy comes from alternative resources. Still, a recent survey of renewable industries in OECD countries[28]—where most of this power is currently generated—revealed important expansionary trends. Modern biomass and biofuel resources, which provided around 89 percent of the OECD's alternative energy in the year 2000, are expected to undergo steady growth in the coming decades. Meanwhile, the energy generated from industrial/municipal waste, geothermal, tidal, and wave projects were collectively providing about 5 percent of the total in 2000, after a decade of healthy expansion. Finally, wind and solar projects expanded more than five-fold in ten years, so that by 2000 these industries were providing about 6 percent of the renewable energy consumed in OECD countries (International Energy Agency 2003).

Wind and solar systems have attracted particular interest as technological improvements have reduced their generating costs around the world. By the year 2000 wind-produced electricity was averaging about 4 cents per kilowatt hour and solar thermal was at about 6 cents in the United States, compared to an average retail cost of 6.8 cents (National Renewable Energy Laboratory 2002). Huge spikes in the cost of conventional electricity in subsequent years made wind and solar projects very competitive in the western United States, at least temporarily. Similar cost improvements have spurred rapid increases in wind projects in Western Europe, with Denmark and Britain taking the lead. Japanese companies, meanwhile, have led the world in commercializing solar thermal and photovoltaic systems. Firms in the United States, South Korea, Taiwan, Mexico, Brazil, and India have also begun manufacturing solar systems for domestic use and export, while China has emerged as a major producer of photovoltaic units. Furthermore, oil companies such as British Petroleum and Royal Dutch Shell have significantly increased their investments in solar systems in the last decade as well.[29] Given these developments, the International Energy Agency (2003) concluded its survey of renewables by asserting that most of these sectors were at last able to compete in wholesale electricity markets in many regions of the OECD.

Even if many modern renewables are approaching price parity, they still generally face the disadvantage that the places and times of peak renewable energy production do not normally match up with commercial demand. Large-scale wind and solar projects, for instance, are most advantageously situated in remote plains and deserts, and their power output can fluctuate considerably. Although more sophisticated electrical grids could accommodate some of these fluctuations, there is still great interest in devising a new system for storing and transporting the energy generated by intermittent and dispersed renewable systems. This is one key reason why so many energy analysts have focused their attention on hydrogen. Another reason, of course, is that supplying renewably produced hydrogen to fuel cells provides one very low-emission option for transportation and stationary power industries.

As with renewables, the commercial hydrogen industry underwent significant expansion in the 1990s. By 2000, the per-unit cost of hydrogen extracted from hydrocarbons was only around two to four times as expensive as gasoline in the United States (assuming the price of gasoline was $1 per gallon).[30] If we recall that oil consumption spread rapidly under the weight of this price differential, it is to be expected that hydrogen sales can undergo similar growth. Advances in solar-heated electrolysis and biological generation systems have also begun to show promise as sources of hydrogen, although renewably produced hydrogen is still very expensive. Meanwhile, studies carried out by the U.S. Institute of Gas Technology have demonstrated that blends of natural gas containing up to 60 percent hydrogen can be used in most existing pipelines, while government tests demonstrated that hydrogen could be safely transported in trucks and used to fuel automobiles. As a recent National Academy of Sciences report concluded, building a hydrogen infrastructure in a country like the United States is now technically and financially feasible.[31]

As occurred at the end of the nineteenth century, the end of the twentieth century has clearly seen the consolidation of a cluster of new fuel and renewable energy industries. The rapid growth of natural gas/methane production makes this resource a strong contender for replacing coal and nuclear power in the coming decades, and eventually supplanting oil as crises intensify in that energy system. Meanwhile, a variety of renewable energy systems are approaching their mass-production phases, just as automobile and electrical systems did one hundred years ago. The question, though, is whether societal dynamics will favor an accelerated expansion of these energy systems. As the next section describes, there is indeed reason to expect that world historical dynamics will foster a shift toward more sustainable energy systems in the coming decades.

World Historical Dynamics and the Shift toward a Sustainable Global Energy System

When viewed from the perspective of great power politics and global energy dynamics, there is a striking similarity between the turn of the nineteenth century and the turn of the twentieth century. During the period 1893–1913, the world-economy went through an economic expansion that corresponded to the last stage of British hegemony. In that era of relative peace and prosperity, one long-established energy system—based on coal—entered into full bloom under the strong promotion of the mature hegemonic state. Meanwhile, new energy systems—based on oil and hydroelectricity—underwent expansions in ascendant states such as the United States and Germany. With burgeoning energy resources flowing from established coal sectors supplemented with new oil and electrical products, consumers and industrialists throughout the core of the world-economy experienced a golden age of economic affluence and resource abundance.

Underneath this veneer of peace and prosperity, however, ominous trends were in motion. The ability of the hegemonic state to contain rising geopolitical tensions was eroded as Britain lost its military, economic, and ideological preeminence. Of crucial importance was the fact that a massive infusion of energy resources into new military systems dramatically increased the lethality of the era's armies. Indeed, it is remarkable how rapidly new military technologies—fueled by new energy systems—spread outward from Europe to North America and even to the Far East. By the end of the first decade of the twentieth century, humanity was therefore poised on the brink of an unprecedented period of chaos and conflict. As we have seen, the global energy system not only played a central role in stimulating this conflict, but it was also deeply transformed during the tumultuous decades that ensued.

If we shift our perspective forward one century, it is possible to identify many of the same dynamics in operation. The last decade of the twentieth century again witnessed a period of economic prosperity that likely corresponds to the final stages of U.S. hegemony. Again, this temporary period of tranquility and affluence allowed another long-established energy system—this time based on oil—to enter into full maturity under the active promotion of the reigning hegemonic state. At the same time, new energy systems—based on natural gas/methane and renewables—came to the verge of commercial viability in many countries. Given this context of relative stability and abundance, it is no surprise that many communities across the world reverted back to energy-intensive lifestyles.

Unfortunately, this period of peace, prosperity, and energy abundance has given way to a new era of global turbulence. On the one hand, the contemporary world faces many of the same hazards that emerged a century earlier. Once again, a weakened hegemonic power is likely to be challenged by rising powers in many regions of the developing world. It is also likely to lash out in preemptive efforts to protect its weakened global position (Harvey 2003). Once again, energy resources are increasing the destructive potential of interstate warfare, while competition for these resources is making conflict more likely. If the hegemonic transition sequence repeats itself as it has in the past, but this time with even more intense energy-driven dynamics, then the world is facing a very bleak future indeed.

Layered on top of these traditional dangers, moreover, is a host of new threats. For instance, materials from nuclear installations are almost certainly spreading to nonstate groups such as al Qaeda. At the same time, the proliferation of other kinds of conventional and unconventional weapons throughout key regions of oil production raises the possibility that massive catastrophes will disrupt the international oil system in the coming years. Even if the worst kinds of disasters can be averted and petroleum production is sustained, the oil system still faces emerging resource constraints.

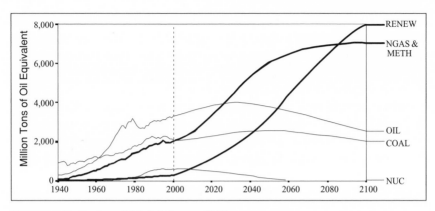

FIGURE 7.3. Projected Global Energy Shifts
Sources: For historical data, see Appendix A.

For the first time in modern history, a major international energy system is approaching a production peak that, within a matter of decades, will put sustained pressure on global prices and availability.

Given these multiple threats, it is imperative that alternatives be developed to ease overreliance on oil. Strong pressures are already emerging, of course, to shift to heavier reliance on unconventional oil, coal, and nuclear power. If these shifts become dominant trends in the global energy system, an unprecedented level of new environmental and security dangers will be unleashed that will further undermine stability and peace across the world.

There is a way out of this interconnected set of energy-driven crises. First, the world community must work to hold coal production to only a modest future increase, while phasing out nuclear power entirely. Meanwhile, we can expect the peak of oil production to occur sometime around 2040, with a slow decline after that. The world community must then prevent a full-fledged shift to heavy oil sands and shale. If the expansion of the natural gas/methane system can be accelerated, replicating the 1940–70 trajectory of oil, pressure to move toward more hazardous resources can be eased for the intermediate term. Meanwhile, fuel cell and hydrogen infrastructures can be fostered within the natural gas/methane system, thereby reducing the environmental impact of gas consumption and setting the stage for a more profound shift to renewable energy resources (see Figure 7.3).

This more fundamental shift can take off in the middle of the twenty-first century if steady growth in modern renewables can be achieved over the coming decades. If the collective energy output of hydroelectricity, biomass, geothermal, wind, solar, and other renewables can replicate the

1960–90 trajectory of natural gas, then these renewables will be making a very solid contribution to the world's energy supplies by the middle of this century. At that point stronger efforts would have to be made to set renewables on a sustained, rapid growth rate—one that more closely resembles the 1940–70 trajectory of oil.

Under this projected growth rate, renewables would provide about 20 percent of all commercial energy by 2050 and 40 percent by 2100. This scenario also allows for aggregate growth in total commercial energy consumption of a little under 2 percent per year up until about 2050, though after that the rate of consumption is assumed to slow somewhat as the world community begins to enact stronger demand-management programs.

The projections described here are quite ambitious, though they are more modest than some scenarios developed by the World Energy Council and the Intergovernmental Panel on Climate Change.[32] It is to be hoped, of course, that even more fundamental reforms can be achieved. Still, the advantage of this approach is that it bases projected trajectories for specific resources on growth curves that have been achieved in earlier historical periods. Therefore, these kinds of shifts are within the realm of historical possibility. The question is, Will dynamics in human society foster these kinds of shifts? A world historical analysis suggests that there are indeed reasons to expect that geopolitical, commercial, and social factors will converge to favor these shifts in the coming years.

If we cast our eyes back across the modern history of the global energy system, it becomes clear that periods of significant transformation in energy industries occur most readily during times of serious global turmoil. The coal system, for instance, was stimulated into rapid regional expansion by the dynamics of intra-European rivalry that occurred at the beginning of the nineteenth century. The oil system, meanwhile, was propelled forward by the even more intense forms of geopolitical conflict that engulfed the world in the mid-twentieth century. Given these precedents, it is likely that new energy systems will undergo rapid expansion as major regional crises and great power conflicts intensify in the early part of the twenty-first century.

Indeed, we can already see state interventions increasing as problems emerge in the maturing oil system. The United States and Britain have taken military action in Iraq to try to reassert control over a crisis-prone Middle East, and partly to defend stability in the international oil system (Harvey 2003). At the same time, however, they have also announced new initiatives to expand fuel cell, hydrogen, and other alternative systems. Indeed, the European Union is undertaking a major effort to accelerate the commercialization of fuel cell and renewables to enhance energy security and reduce greenhouse gas emissions. Japan is engaged in similar large-scale efforts, while China, India, Mexico, Chile, and a number of other developing nations are adopting policies designed to increase the use of natural gas and renewables. In fact, the Chinese government recently announced that it would impose fuel economy and emissions standards on new cars and trucks that will be stricter than those in the United States.[33] States across the world are clearly engaged in important efforts to spur reforms in their domestic energy sectors.

There is another factor that is likely to enhance the effectiveness of these political interventions in global energy industries. Future energy transitions will be facilitated by multilateral agencies that can assist in setting common agendas and coordinating policies at the international level. Although organizations such as the World Bank and the International Energy Agency have historically directed most of their support toward conventional energy industries, there are indications that these institutions are in the process of modifying their priorities. Pressure from nongovernmental organizations and the European Union has caused the World Bank to commit itself to increasing funding for environmentally sustainable energy projects. Multilateral institutions are also assisting in efforts to develop mechanisms for transferring clean energy technologies to the developing world.[34] Given this new level of international coordination, it is conceivable that a transition to natural gas and renewable energy systems can take place faster than earlier global energy shifts.

Even if the United States loses its lead in these innovative efforts because of the entrenched power of its oil, coal, and nuclear power industries, this does not necessarily mean that a global shift toward natural gas and renewables will be stopped. Remember that France and Germany remained wedded to coal long after oil had emerged as the era's most important energy industry, to the detriment of those countries but to the advantage of others. A world historical perspective reminds us that what is of most importance is not the behavior of one single state—even if it is a declining hegemon—but how dynamics of competition impact the behavior of many states.

In this regard, what is probably most crucial is how ascendant nations like China, India, South Korea, and Brazil respond to contemporary energy challenges. If these nations tap their increasingly skilled working classes to mass-produce fuel cells, wind turbines, solar panels, and components for the hydrogen infrastructure, then the transition to a renewable energy system can be greatly accelerated. Indeed, the most successful business model for new energy systems will undoubtedly involve mass-producing the components in moderate- to low-wage countries, and then selling these products to consumers in pollution-choked cities and affluent communities across the world. Adoption of this globally oriented manufacturing and marketing strategy would reflect an important innovation in the corporate history of the global energy system. It would also bring costs of new energy systems down significantly, though it would limit the number of high-wage jobs created by these emerging industries.

Signs that this business model is already being implemented are not hard to find. In 2004, for instance, Hyundai debuted a fuel cell powered SUV that is manufactured in South Korea and aimed at international markets. Meanwhile, a Canadian/Chinese venture finalized plans to open a manufacturing plant in Shanghai designed to produce twenty thousand fuel cells per year. These fuel cells, which can be stacked into arrays of various sizes to replace components like diesel generators or oil burners, will be sold in China and overseas.[35] Again, while these cells are likely to be relatively inexpensive—which will help accelerate the next global energy shift—workers in these plants will face very different working conditions than auto, oil, or utility workers experienced decades ago in the developed world.

Just as we should not expect emerging renewable energy industries to generate utopic working conditions, we should also not expect these industries to be driven forward by decentralized, locally based companies. Advocates of renewable technologies have long claimed that the expansion of wind, solar, and fuel cell systems would foster many small, nonhierarchical firms. Of course, renewable sectors will not be nearly as centralized as nuclear power industries. But, while many small firms may survive in certain areas, alternative energy industries are already attracting the interest of multinational petroleum and utility corporations. Recall that the coal system was populated by many small, regionally based firms, while the oil-based energy system was dominated by a few, internationally oriented corporations. Today, new energy systems such as the fuel cell are being developed by joint ventures that link innovative engineering firms with multinational corporations from automotive, electrical, and oil industries. Government agencies in North America, Western Europe, and Asia are also providing crucial support to these joint ventures. By combining resources from multiple realms, these joint ventures appear to be well situated to push the commercialization of new energy technologies forward.

Partnerships between private companies, national governments, and international lending institutions are also going to have to play a central role in mobilizing the capital required to build new energy infrastructures.

The financial costs of building new transcontinental pipelines, electrical grids, and generation/distribution systems will certainly be substantial.[36] Yet, there are many reasons to expect that these challenges can be met. Remember that huge investments were made in the nineteenth and twentieth centuries in coal and oil infrastructures, even during times of economic turmoil and military conflict. Construction projects in the twenty-first century should likewise be able to forge ahead, even in the midst of significant global turbulence. Future projects will be aided by the existence of more sophisticated partnerships, international capital markets, and improved engineering practices than existed in previous centuries. Finally, it is worth noting that these infrastructure projects can provide a much-needed outlet for excess capital. As Brenner (2002) points out, advanced capitalist economies are plagued by overcapacity in established manufacturing sectors. If a new wave of investments in infrastructures designed to bring natural gas, methane, and renewable energy products to consumers can be orchestrated, these economic bottlenecks can be eased while more sustainable energy systems are expanded.

Just as there are reasons to expect political and commercial dynamics to foster a shift toward new energy systems, there is also evidence suggesting that social dynamics will help accelerate such a shift. Perhaps the most powerfully transformative pressure will come from rising social conflict in centers of oil production. Indeed, the early twenty-first century is witnessing unparalleled levels of social unrest in the Middle East. The 2003 Iraq war and subsequent occupation of that country by the United States and its allies removed the tyrannical regime of Saddam Hussein, but it has also galvanized anti-Western radicals and may leave an enduring civil war in its wake. Meanwhile, the Israeli/Palestinian conflict continues to reach new levels of intensity, and evidence suggests that both the Iraqi and Israeli/Palestinian conflicts are radicalizing a new generation of youth in the region, which could be a development of far-reaching consequence.[37]

The conservative political and social structures of Saudi Arabian society have already fostered the emergence of a cadre of militants who play an important role in al Qaeda. With new recruits, these armed groups will increase their ability to carry out attacks in Saudi Arabia and neighboring countries. This emerging militant threat, when combined with the insular regimes and social polarization found throughout the region, raises the possibility that other oil-exporting nations will undergo the kind of instability that gripped Iran in the late 1970s.[38] As the Iranian Revolution demonstrated, social conflict in oil-rich countries can translate rapidly into price spikes in international petroleum markets and general financial turmoil around the world. As social crises grip the Middle East, therefore, added pressure will be generated for an accelerated shift to new energy resources.

As that pressure mounts, it will be vitally important that social movements work to forestall a wholesale shift to nuclear power, unconventional oil, and coal. As we have seen, antinuclear activists played a key role in halting the global expansion of this energy system in the 1980s. The current era demands a revival of these broad-based mobilizations, this time to prevent a resurgence of the industry. Though there are powerful interests behind the nuclear power industry, the economics of the industry remain unfavorable. Moreover, heightened public concerns about recent nuclear accidents and terrorist attacks can give added weight to antinuclear campaigns. In short, the prospects for success in preventing a large-scale resurgence of civilian nuclear power appear promising, provided that concerned citizens engage the battle.

Preventing shifts toward greater reliance on unconventional oil and coal resources poses significant challenges. There are powerful interest groups in countries like Canada and China that advocate the development of heavy oil and shale deposits, while coal production is again on the rise in every region of the world. Moreover, there are important divisions between social movements over the advantages and drawbacks of these industries. For instance, labor unions like the AFL-CIO have rejected the Kyoto Treaty, and this federation joins with many unions across the world in advocating an expansion of oil and coal production.[39]

On the other hand, some major unions have recognized the threats posed by global climate change and have committed themselves to supporting a transition toward sustainable energy systems. The twenty million member strong International Federation of Chemical, Energy, Mine, and General Workers' Union (ICEM), for instance, has called for a "just transition" that helps working class communities negotiate the shift away from heavily polluting industries and toward new energy systems.[40]

At the same time, new labor struggles are emerging in countries like China, where coal mining takes a disastrous toll on workers. Indeed, somewhere between six thousand and seven thousand people lost their lives in Chinese coal mines in 2003 alone. While it is extremely difficult to carry out traditional union campaigns in China, there are new community- and party-based campaigns under way to try to reign in the worst of these mines, and to reduce the immense pollution being generated by coal consumption across the country.[41]

There is ample opportunity for labor and environmental groups to coordinate their efforts to accelerate a socially just transition toward natural gas, methane, and renewable energy systems. Indigenous rights groups can also be incorporated into this broad campaign. Indeed, some of the fiercest resistance to expanding oil and mining operations is now coming from indigenous peoples in Latin America, Africa, and Asia. With the assistance of international organizations like the Rainforest Action Group and Amazon Watch, indigenous peoples have raised global awareness of their plight and engaged in legal campaigns to protect their homelands. Some have even used kidnapping and armed resistance to try to keep oil and mining companies at bay. If human rights and environmental organizations headquartered in the global North can increase their support for these indigenous communities, the continued expansion of oil and coal extraction may be more effectively held in check until alternatives can be developed.[42]

Meanwhile, environmental groups and citizens in the global North have a particular responsibility to reform energy consumption patterns in their own countries. While new technologies provide opportunities for reducing some pollution, there is no escaping the fact that people in the advanced industrial world will have to shift toward less energy-intensive lifestyles.

This problem is particularly acute in the United States, where a long history of resource abundance has fostered the creation of a high-energy society (Nye 1999). There have been some efforts within the United States to raise public awareness about this problem of overconsumption. For instance, media campaigns undertaken by mainstream groups and arson attacks carried out by the Earth Liberation Front have raised public awareness about the problematic impacts of SUVs. A broad-based coalition of mainstream environmental groups has also tried to prevent the Bush Administration from unraveling decades of regulatory reforms. However, much more could be done by citizens in the United States and in other affluent countries to address the problem of overconsumption. It is an urgent task, because the inequalities in energy use that have intensified since the Second World War must be eased if major global tensions are to be avoided. As many as 1.6 billion people throughout the developing world still have no access to electricity. Most of them live in countries located in tropical and subtropical areas, where power grids are difficult to construct but where renewable energy resources are abundant. Given these characteristics, there are considerable advantages to expanding small-scale hydroelectric, wind, solar thermal, and photovoltaic systems throughout the developing world.[43] Companies headquartered in developed and developing countries are on the verge of mass-producing these systems. And, under pressure from activist groups, multilateral development institutions like the World Bank are finally lending their expertise to efforts to accelerate the international diffusion of these systems. The world policy environment therefore appears set at last to foster the expansion of new energy systems. Although it is impossible to predict how quickly new energy technologies can spread, it nevertheless appears that they are in a strong position to begin processes of rapid diffusion in the coming decades.

There are inherent uncertainties in the manner in which each of the political, commercial, and social dynamics that have been examined in this study will evolve. Even more unclear is the way in which they will interact to prevent or promote a shift toward a more sustainable energy system. Scenarios can of course be constructed in which such a shift is derailed. Of principal concern is the possibility that escalating tensions between major world powers, themselves at least in part stimulated by conflicts over energy resources, might generate open warfare in the coming decades.[44] Under the impact of such destructive geopolitical conditions, the commercial and social dynamics promoting the expansion of new energy systems would evaporate. Of course, in this kind of scenario there would be so many other dramatic crises convulsing the world-system that a stalled shift toward renewable energy systems would represent a minor concern.

Barring the onset of global warfare, however, there are strong reasons to expect that a global energy shift can be achieved in the twenty-first century. If political and social conflict continue to be endemic in the Middle East, then the price of oil will rise and import-dependent states will have reason to increase support for domestically based alternative energy technologies. Private companies and consumers will also be encouraged by these market pressures to turn toward new energy systems. Under such a scenario, even relatively moderate environmental pressures on conventional energy industries could be sufficient to prompt a more rapid diffusion of alternative energy systems.

The contemporary global energy crisis poses one of the most serious policy challenges of the current era. Growing volumes of energy must be provided for development projects in Asia, Africa, and Latin America, during a time when conventional energy paths no longer appear to be sustainable. Given these dilemmas, many have come to the conclusion that fundamental change is impossible, and that an environmental catastrophe of global proportions will eventually undermine the integrity of the capitalist world-economy. The truth of the matter, however, is that there are paths that can be followed toward a more sustainable global energy future. Rapid and far-reaching global energy shifts have occurred in the past, and they can take place again. As perhaps never before, the technological options appear to be open. The task now is to fortify the actions of states, companies, and communities to support the transition to a more environmentally sustainable and equitable energy system in this century.

Appendix A
Sources and Methods Used
to Compile Energy Data

The analyses undertaken in this study are based on data covering coal, petroleum, natural gas, nuclear, hydroelectric, and alternative energy industries for the period 1800–2000. The following sources were drawn upon for the production, trade, and consumption data: for the period 1800–1949: Etemad and Luciani (1991), Great Britain Central Statistical Office (1996), Mitchell (1984, 1988), and the U.S. Department of Commerce (1975), and for the period 1950–2000: *The United Nations Energy Statistics Database*, provided in the annual volumes published by the United Nations, entitled *Energy Statistics Yearbooks*, and supplemented by computerized files provided by the United Nations Energy Statistics Unit. Additional trade and consumption data for the years 1925–49 were taken from United Nations (1952), from Darmstadter et al. (1971), and from volumes of historical statistics compiled by Brian Mitchell (1982, 1983, 1984, 1988, 1992, 1993). Meanwhile, data on traditional biomass and modern alternative energy sectors were drawn from reports provided by the World Energy Council, the International Energy Agency, and the U.S. Energy Information Administration (see specific citations in the text).

Since particular energy resources have different quantities of useful energy per volume unit, it is necessary to convert each distinct resource into comparable units before data aggregation or comparison is undertaken. The United Nations data source, as well as various International Energy Agency publications, provides information on internationally recognized factors used to convert different energy resources into comparable units. These comparable units can be reported as *tons of coal equivalent, tons of oil equivalent,* or *terajoules.*

United Nations publications presented data in coal equivalencies until 1976, and then began providing information in all three comparable units in subsequent publications. The International Energy Agency publications generally provide data in oil equivalencies, as do most private energy research publications (such as *The British Petroleum Survey of Energy Resources*). For this study, I have chosen to convert volume units of measurement into *oil equivalencies* for the purpose of aggregating and comparing energy production across resource types. General conversion factors provided in the United Nations *Energy Statistics Yearbooks* were most frequently used. However, in some cases the more specific conversion factors for particular countries provided in the series *Energy Statistics & Balances of OECD Countries* and *Energy Statistics & Balances of Non-OECD Countries* (International Energy Agency) were used.

Reliability checks were also carried out on the production, trade, and consumption data once the conversions had been done. Comparisons of data presented in Etemad and Luciani (1991) data, the United Nations *Energy Statistics Yearbooks*, the International Energy Agency energy publications, the U.S. Energy Information Administration's *Annual Energy Review*, and the *British Petroleum Survey of Energy Resources* reveal a generally high level of reliability. This does not necessarily mean that the data presented in these sources is correct—merely that errors are reflected in all the recognized energy data sets. The data used in this study inevitably contain the same measurement errors. Of course, responsibility for any errors in the data lies entirely with the author of this study.

The United Nations data on production, trade, and consumption begins systematic coverage around 1925, though for certain countries there are gaps. Due to the data collection efforts of Etemad and Luciani (1991), information on commercial energy production for most countries is quite complete, even for the nineteenth century. However, information on trade and consumption generally begins in 1925.

The following is a list of the years that coverage on energy production begins for each country in the energy data set, ordered by year coverage begins: Austria (1800), Belgium (1800), France (1800), Germany (1800), Poland (1800), the United Kingdom (1800), the United States (1800), Ireland (1819), Hungary (1828), Sweden (1834), Italy (1846), Australia (1848), Russia (1850), Canada (1851), India (1851), Spain (1853), Romania (1858), Czechoslovakia (1860), the Netherlands (1860), Turkey (1863), Portugal (1864), Switzerland (1866), Japan (1869), New Zealand (1871), Greece (1875), Denmark (1876), Bulgaria (1879), Yugoslavia (1880), Mexico (1881), Peru (1884), South Africa (1884), Chile (1888), Indonesia (1889), Malaysia (1890), China (1894), Taiwan (1896), Brazil (1898), Norway (1900), Zimbabwe (1900), Luxemburg (1901), Venezuela (1904), Iran (1906), the Philippines (1907), Argentina (1908), Trinidad (1909), Uruguay (1910), Egypt (1911), Guatemala (1912), North Korea (1912), Nigeria (1915), Madagascar (1916), Syria (1916), Algeria (1917), Ecuador (1917), the Soviet Union (1917), Albania (1918), Burma (1918), Finland (1918), Bolivia (1919), El Salvador (1919), Jordan (1919), Kenya (1919), Colombia (1920), Honduras (1920), Paraguay (1920), Zaire (1920), Costa Rica (1921), the Dominican Republic (1921), Hong Kong (1921), Morocco (1921), Congo (1922), South Korea (1922), Thailand (1922), Indochina (1923), Jamaica (1923), Tunisia (1923), Cuba (1924), French West Africa (1924), Ghana (1924), Iceland (1924), Nicaragua (1924), Sudan (1924), Afghanistan (1925), Angola (1925), Bermuda (1925), Cameroon (1925), Cape Verde (1925), Cyprus (1925), Fiji (1925), French Equatorial Africa (1925), Gambia (1925), Guadeloupe (1925), Guyana (1925), Haiti (1925), Malawi (1925), Malta (1925), Martinique (1925), Mauritius (1925), Mozambique (1925), Reunion (1925), Rhodesia (1925), Sarawak (1925), Sierra Leone (1925), Sri Lanka (1925),

Tanganyika (1925), Togo (1925), Zanzibar (1925), New Caledonia (1926), Panama (1926), Papua New Guinea (1926), Iraq (1927), the Netherlands Antilles (1928), Barbados (1928), Guinea (1928), the Solomon Islands (1928), Puerto Rico (1929), Brunei (1931), Saudi Arabia (1931), Liberia (1932), Somalia (1932), Surinam (1932), Bahrain (1933), Lebanon (1933), Western Sahara (1934), Kuwait (1935), Ethiopia (1936), Qatar (1936), Mongolia (1938), Greenland (1943), East Germany (1946), Gibraltar (1946), Libya (1946), West Germany (1946), Yemen (1946), Democratic Yemen (1947), Mali (1947), Pakistan (1947), Singapore (1947), Israel (1949), the Ivory Coast (1949), Dominica (1950), Equatorial Guinea (1950), Nepal (1950), Rwanda (1950), St. Vincent (1950), Swaziland (1950), Zambia (1950), Kampuchea (1951), Uganda (1952), Burkina Faso (1953), Samoa (1953), Central African Republic (1954), Laos (1954), Macau (1954), South Vietnam (1954), Gabon (1955), North Vietnam (1955), United Arab Emirates (1956), Burundi (1958), Djibouti (1958), Niger (1959), Guam (1960), Mauritania (1960), South African Customs Unit (1960), Sao Tome (1960), Oman (1961), Senegal (1961), Tanzania (1961), the Bahamas (1962), Vanuatu (1962), Antigua (1963), American Samoa (1963), the Cayman Islands (1964), Grenada (1964), Bangladesh (1967), the British Virgin Islands (1967), Guinea Bissau (1967), Belize (1968), Benin (1968), Bhutan (1968), the Comoros Islands (1970), French Polynesia (1970), Vietnam (1970), Maldives (1971), Nauru (1971), Botswana (1973), Chad (1973), French Guiana (1974), Seychelles (1979), Tonga (1979), Montserrat (1983), Aruba (1985), Pacific Islands (1985), Kiribati (1990), Liechtenstein (1990), Monaco (1990), Namibia (1990), Sabah (1990), San Marino (1990), St. Helena (1990), St. Kitts (1990), St. Lucia (1991), Armenia (1992), Azerbaijan (1992), Belarus (1992), Bosnia (1992), Croatia (1992), the Czech Republic (1992), Estonia (1992), Georgia (1992), Kazakhstan (1992), Kyrgyzstan (1992), Latvia (1992), Lesotho (1992), Lithuania (1992), Macedonia (1992), Moldova (1992), Slovakia (1992), Slovenia (1992), Tajikistan (1992), Turkmenistan (1992), Ukraine (1992), Uzbekistan (1992), and Niue (1994).

Where missing data have been estimated, the method of linear interpolations between data points has been used. This method is widely used, since national patterns in energy production, trade, and consumption generally follow smooth trajectories. Given missing data problems during the years 1940–45, the series on consumption was left as missing for this period.

Appendix B
Glossary of Petroleum
Company Names

The names of some major petroleum companies have changed over the history of the industry. In this study, oil companies are referred to by their names at the time specific events took place. The following glossary provides a list of major changes in corporate names that occur in the text.

Current Name	Historical Names Used in Text
Exxon/Mobil	Standard Oil of New Jersey, Jersey Standard, Exxon, Standard Oil of New York, Socony, Mobil
Chevron/Texaco	Standard Oil of California, Texaco
BP/Amoco/ARCO	Anglo Persian Oil Company, British Petroleum, Amoco, ARCO
Royal Dutch Shell	Royal Dutch, Shell Oil Company
Total/Fina/ELF	Cie Francaise des Petroles, CFP, ELF

Appendix C
Sources of Strike Data in
Energy Industries

This study utilizes a variety of data on strikes to trace out patterns of labor unrest in energy industries. In general, the data can be divided into two categories: official strike series (available for Great Britain and the United States) and a database on newspaper mentions of labor unrest (available for most countries of the world). The following is a description of the data sources used in this study.

For the United States, official information on the number of strikes, number of workers involved, and number of days idled in coal and petroleum industries came from the following sources: for the period 1899–1955: U.S. Department of Commerce, *Statistical Abstract of the United States* (various volumes), and for the period 1955–80: U.S. Bureau of Labor Statistics, *Analysis of Work Stoppages* (annual bulletins). Post-1980 official data on U.S. coal strikes is no longer available in a comparable form. For Great Britain, official information on strikes came from the following sources: Milner and Metcalf (1991) and Great Britain Central Statistical Office (1996).

While some official information on strikes also exists for Australia, Canada, France, and Japan, these core-focused data series are inadequate for tracing out global waves of strike activity in energy industries. For this reason, a second source of data, the World Labor Group Database on Labor Unrest, was used to map out world-scale patterns of labor unrest.

The World Labor Group Database was compiled from the indexes of the *Times of London* and the *New York Times*. It includes all incidents of labor unrest (e.g., strikes, demonstrations, factory occupations) throughout the world reported by these two sources from 1905 to 1990. The database also contains mentions of labor unrest from the *New York Times* for the period 1870–1905. Reports are coded by action type, location, and date. Extensive reliability studies have been carried out on the database, showing that it can correctly identify major waves of unrest in specific countries (see Silver 2003). The database has also been used to identify waves of unrest in specific industries, such as automobiles, mining, textiles, and railroads. Using this data set, the inadequacies of official strike series may be partially overcome and truly global patterns of unrest can be mapped. Special thanks is extended to Professor Beverly Silver (Department of Sociology, Johns Hopkins University)—the key researcher involved in the construction of this database—for permission to use World Labor Group data in this study.

Notes

Chapter One

1. See Jevons (1865), Arnot (1866), and Freese (2003) for analyses of coal dilemmas that existed in Britain in the late nineteenth century.

2. For two particularly influential reports articulating these positions, see Donnelly (2000) and National Energy Policy Development Group (2001).

3. On the rise of social tensions, see Okruhlik (1999); on geopolitical dangers, see Center for Strategic and International Studies (2001); and on environmental threats, see Intergovernmental Panel on Climate Change (2001).

4. In a true sense, all energy resources have their origins in solar energy. However, because they present themselves in distinct forms in the natural world, it makes practical sense to discuss them as separate forms of energy.

5. See Appendix A for a discussion of the sources and methods underlying the statistical data on energy used in this study.

6. See Appendix A for more information on this procedure.

7. A sampling of leading scholars in this research tradition include McNeill (1982), Braudel (1992), Wallerstein (1974), Tilly (1984), Chase-Dunn (1989), and Arrighi and Silver (1999). Because some of these scholars do not identify themselves as world-systems researchers, I have chosen to use the more general term of *world historical research* to encapsulate this broader community of analysts.

8. For complementary, if somewhat differing, discussions of the historical dynamics of interstate competition, see Choucri and North (1975), Wallerstein (1984), Kennedy (1987), Goldstein (1988), Modelski and Thompson (1996), and Chase-Dunn and Podobnik (1999).

9. On the history of wood use, see Perlin (1989) and Chew (1995: 203–4).

10. Readers should also consult the following sources for more details on state-energy interactions: Ikenberry (1988), Jones (1981), Kapstein (1990), Keohane (1984), Krasner (1978), Nash (1968), Painter (1986), Samuels (1987), Stivers (1982), and Vernon (1983).

11. See Bunker and O'Hearn (1993) and Bunker and Ciccantell (1999).

12. A world historical interpretation of the commercial dynamics of capitalist development has been advanced by such researchers as Schumpeter (1949), Hymer (1979), and Arrighi (1994).

13. On labor unrest see Dix (1988), Davidson (1988), Feldman and Tenfelde (1990), and Colley (1997); on nationalist movements see Nore (1980) and Brown (1993); and on environmental and indigenous rights movements see Rudig (1990), Nilsson and Johanssen (1994), and Gedicks (1995).

14. Perhaps the most influential book on energy matters, Yergin (1991), exemplifies this underexamination of dynamics of social conflict (though his examination of political and commercial dynamics is outstanding).

15. It should be emphasized that the intention here is not to argue that a hegemonic world order is solely determined by resource advantages. While gaining access to secure and affordable resources is certainly a key prerequisite for the establishment of hegemony, it is only one of a number of challenges that must be addressed by a rising great power.

16. Examples of world-systems scholars who steadfastly maintained that the United States was in decline during the 1990s are numerous, and include Wallerstein (1992), Arrighi (1994), and Boswell and Chase-Dunn (2000).

17. See Pratt (1981), Perelman (1981), Criqui (1994), Kemp (1994), and Martin (1996) for other analyses of large-scale energy shifts that emphasize the impact of human institutions.

18. See Cottrell (1955) and Tainter (1990) for discussions of the relationship between social conflict and energy scarcity in preindustrial eras.

Chapter Two

1. The historical discussion presented in the next two paragraphs draws on the following sources: Eavenson (1939), Hartwell (1960), Needham (1964), Shepherd (1993), and Kopp (1995).

2. See Thomas (1986), Perlin (1989), Fouquet and Pearson (1998), and Sieferle (2001) for more information on deforestation pressures in Europe.

3. Cf. Marx (1967), Weber (1947), Polanyi (1957), and Wallerstein (1979) for more complete discussions of this transition. My interpretation of the significance of the Treaty of Westphalia, discussed in the following paragraphs, is based on the analyses of Arrighi (1994).

4. As discussed in the next chapter, the solidification of the third systemic dynamic—that of social unrest—would have to await the consolidation of labor forces in concentrated communities, as well as the emergence of traditions of worker organizations.

5. Information presented in the next two paragraphs on the development of steam technologies was drawn primarily from the following sources: Ferguson (1967), Harre (1984), and Smil (1999).

6. Jeremy (1977); Stein (1984).

7. Redlich (1944); Robinson (1974).

8. Thomas (1986); Fouquet and Pearson (1998).

9. The following discussion draws on Dickson (1967), Cain and Hopkins (1980), and Brown (1991).

10. Information in this paragraph is drawn from the following sources: Spring (1951), Cairncross (1953: 98), Pollard (1965), Mathias (1969: 244), McCahill (1976), Cromar (1977), and Flinn and Fogleman (1984: 209).

11. Devine (1976) discusses the Scottish case, while Clemens (1976) provides information on the Liverpool case.

12. See Pollard (1978, 1980) and Burt (1991) for discussions of the mobilization of mine labor.

13. For more on the labor process and communities characteristic of early British mining, see Dennis, Henriques, and Slaughter (1956), Gilbert (1992), and Williams (1998).

14. See Cromar (1977) and Flinn and Fogleman (1984) for details on early working conditions in British mines.

15. Cf. McNeill (1982: 177), Henderson (1975: 57), and Parnell (1994: 12).

16. On the use of coal-based systems in the Crimean War, see Kipp (1976: 37) and Lambert (1996: 159); on the Franco-Prussian War, see Heinze and Kill (1988) and Heaton (1967: 511); and on the development of steam-powered naval capacity, see McNeill (1982) and Griffiths (1997).

17. Cooling (1979: 215); Mann (1988: 172); Trebilcock (1993: 567).

18. Information in this paragraph was drawn from the following sources: Reid (1985), Michel (1990), and Lewis (1993).

19. For Belgian coal mining, see Hentenryk and Puissant (1990); for Germany, see Henderson (1975), Showalter (1975), and Parnell (1994); and for France, see Fohlen (1978), Reid (1985), and Lewis (1993).

20. Robinson (1972); Dumett (1975); Sullivan (1983).

21. Headrick (1988: 32); Lambert (1996: 156).

22. Pollard and Robertson (1979); Headrick (1981); Lambert (1996).

23. Crouzet (1990: 431).

24. Cf. Cairncross (1953), Bloomfield (1968), Dunning and Archer (1993), Wilkins (1993), and Cameron (1997).

25. Taylor (1968: 61); Crouzet (1990: 436).

26. Hunter (1985: 416); Palladino (1990: 39).

27. Cf. Goodrich (1977), Seltzer (1985), Dix (1988), Bowman (1989), and Long (1989).

28. Wynn (1992); Stearns (1993); Siegelbaum and Walkowitz (1995).

29. Minami (1977); Broadbridge (1989); Kazuo (1997).

30. Kimura (1995); Yasuba (1996).

31. Wright (1981); Yuru (1994).

32. Foster (1999); Moore (2003).

Chapter Three

1. Landes (1969: 231); Ville (1990).

2. Cf. Michel (1990), Hickey (1985), Dix (1988), Magraw (1992), Parnell (1994), and Shorter and Tilly (1974).

3. See Arrighi, Hopkins, and Wallerstein (1989) for a more detailed discussion of the impact of these revolutions.

4. Hobsbawm (1964); Price (1988); Hilden (1993).

5. The strike-prone nature of coal mining has been documented in many studies, including those by Kerr and Siegel (1964), Shorter and Tilly (1974), Edwards (1981), and Feldman and Tenfelde (1990).

6. For discussions of the concept of workplace bargaining power, see Perrone (1983, 1984), Arrighi and Silver (1984), Wallace, Griffin, and Rubin (1989), and Silver (2003).

7. Church (1986: 714); Fine (1990).

8. Geary (1985: 6); Milner and Metcalf (1991).

9. Reid (1985); Michel (1990); Magraw (1992).

10. Geary (1981: 105); Geary (1989: 115); Hickey (1985: 170); Parnell (1994: 22); Rimlinger (1989: 580).

11. Hentenryk and Puissant (1990: 228); Hilden (1993: 109).

12. Montgomery (1980: 90); Edwards (1981: 274); Seltzer (1985: 261); Brody (1990: 86); and Blatz (1994).

13. Information on Canadian strikes comes from Jamieson (1968), Heron (1989), and Kealey (1995).

14. Information on the early history of oil can be found in Forbes (1958) and Owen (1975).

15. Johnson (1967: 666); Rae (1967: 347).

16. Moran (1987: 579); Isser (1996: 4).

17. Over time, oil companies have been divided and entered into mergers. As a result, their names have changed substantially. Consult Appendix B for a list of evolving corporate names in the international oil industry.

18. Debeir, Deleage, and Hemery (1991: 114); Yergin (1991: 51).

19. Aliyev (1994).

20. Henriques (1960: 76); Howarth (1997: 36).

21. Marcus Samuel, a British citizen, tried with little success to become the British Navy's main oil supplier in the 1890s. When Shell merged with Royal Dutch Oil, the company became even more suspect in the eyes of British authorities and was often treated with open hostility (Howarth 1997: 103).

22. Based on the declared assets of the Standard companies in 1912, if the dissolution had not occurred the Standard Oil Trust would have been ranked as the world's largest enterprise. The information on company rankings in 1912 comes from Schmitz (1995: 87).

23. U.S. Bureau of Labor Statistics (1950); Werner (1963).

24. Hunter and Bryant (1991).

25. Church (1971); U.S. Bureau of Labor Statistics (1979: 34).

26. McLaughlin (1954); Hard and Jamison (1997).

27. Machlup and Penrose (1950) provide a useful discussion of the spread of patent law in this period.

28. The following discussion of the development of the internal combustion engine draws on these sources: Laux (1963), Barker (1985), and Cummins (1989).

29. Henning and Trace (1975: 353); Ville (1990: 51).

30. Simonson (1960: 362); Fearon (1985: 22); McFarland (1990).

31. Barker (1985); Sabel and Zeitlin (1985).

32. McLaughlin (1954); Aldcroft (1964).

33. Laux (1963); Foreman-Peck (1982).

34. Motor vehicle production data come from the 1911 edition of the Encyclopedia of Social Science, as well as Foreman-Peck (1982).

35. Simonson (1960); Myerscough (1985); Gross (1990).

36. Sharlin (1967); Cantor, Gooding, and James (1996); Buchanan (2003).

37. See Hughes (1983), Hirsh (1990), and Nye (1990) for analyses of this battle of the systems.

38. Kocka (1981); Siemens (2002).

39. Brittain (1974); Nye (1990).

40. For trolley information in the United States, see Schurr et al. (1990) and Nye (1990: 96); for Europe, see Buchanan (2003); and for other regions, see Minami (1977) and Clark (1990).

41. Devine (1983); Neufeld (1987); Lee (1988: 359); Emmons (1993).

42. Marsh (1928); Nye (1990: 208).

43. U.K. electricity consumption data come from Mitchell (1988), while information on all other countries comes from Etemad and Luciani (1991).

44. Sharlin (1967); Devine (1983); Hunter and Bryant (1991).

45. Data from U.S. Department of Labor (2000).

46. See Paris (1991) for details on the use of oil-powered vehicles and airplanes in colonial wars.

47. On the shift to oil-powered ships in Europe, see Ville (1990: 60), Herwig (1991), and Howe (1996: 268); in the United States, see DeNovo (1955) and Griffiths (1997: 155). On the characteristics of the Dreadnought, see Fairbanks (1991).

48. Henderson (1948); Nowell (1994).

Chapter Four

1. Voskuil (1942); Cohn (1993: 16); Wilson and Prior (2001).

2. Grieve and Newman (1936); Barrie (1961).

3. Henning and Trace (1975); Foreman-Peck (1982); Fearon (1985); Myerscough (1985).

4. Feldman (1966: 459); Deese (1981); Lipschutz (1989); Wilson and Prior (2001).

5. Jensen (1968); Hughes (1983: 286).

6. Royal Commission (1926: 218); Taylor (1968: 51); Arnot (1975); Supple (1987: 102).

7. Michel (1990: 283); Cohn (1993).

8. Hickey (1985: 4); Weisbrod (1990: 159); Parnell (1994: 41); Hentenryk and Puissant (1990: 245).

9. Peterson (1938: 1055); Bowman (1989: 134); Brody (1990: 93); Dulles and Dubofsky (1993: 218); Heron (1989: 59).

10. Melby (1981); Bromley (1991).

11. DeNovo (1956); Krasner (1978); Ikenberry (1988).

12. Stivers (1982); Hourani (1991); Said (1994).

13. Penrose (1968); Roncaglia (1983).

14. All these comparisons are done on the basis of calorific values. See Bairoch (1993) for a summary of energy price data.

15. See Schurr and Netschert (1960: 100–3), Williamson et al. (1963: 181), Dix (1988: 186–7), and Nowell (1994:75).

16. Paxson (1946); Overy (1973); Yago (1983); Hugill (1993).

17. Chandler (1962); Isser (1996: 5).

18. See Schurr and Netschert (1960:103), Williamson et al. (1963: 208), Giebelhaus (1982: 99–100), and Howarth (1997:120) for discussions of early market promotion efforts carried out by these oil companies.

19. Penrose (1993: 58); Vernon (1993: 72).

20. Harvey (1982: 181) points out the general difficulties in constructing reliable measures of profit rates, while Williamson et al. (1963: 334) and Philip (1982: 17)

outline some of the limitations of publicly available data on profit rates in the oil industry specifically.

21. Data on oil company profit rates come from the following sources: Williamson et al. (1963: 334, 563), Gibb and Knowlton (1956: 504), Ferrier (1982: 232), Bamberg (1994: 23), and Philip (1982: 12).

22. Data on coal company profit rates come from these sources: Royal Commission (1926: 218), Cohn (1993: 17), Parnell (1994: 49), and Seltzer (1985: 35–6).

23. Williamson et al. (1963: 329); Sampson (1975); Turner (1983).

24. Cf. Chandler (1962, 1990), Harvey (1982: 146), and Schoenberger (1997) on the general difficulties, but competitive necessity, of carrying out processes of organizational decentralization.

25. For discussions of the early management characteristics of these firms and the difficult processes of change discussed in the next paragraph, see the following: Standard Oil: Gibb and Knowlton (1956: 571, 606) and Chandler (1962); Shell Oil: Henriques (1960: 64–5) and Howarth (1997: 212); Royal Dutch: Yergin (1991: 369); Anglo Persia: Bamberg (1994: 521).

26. Banks (1985: 99); McKern (1996: 333).

27. See Pollard (1994) for a good summary of the criticisms leveled at British entrepreneurs, and Chandler (1972) for the U.S. case.

28. On the effects of ruinous competition in British coal, see Taylor (1968: 51–61), Payne (1978: 226), and Boyns and Wale (1996: 62). For the United States, see Seltzer (1985: 35–6) and Bowman (1989).

29. Goodrich (1977) provides a particularly good account of the changes in the labor process of underground coal mining brought about by mechanization.

30. Dix (1988); Blatz (1994); Winterton (1994).

31. For data on mechanization, see Supple (1987: 31) on Britain; Holter (1992) on France; Parnell (1994: 49) on Germany; and Dix (1988: 217) on the United States.

32. Veenstra and Fritz (1936: 111); Murphy and Spittal (1945: 628).

33. Venn (1986: 85); Wilkins (1993: 37).

34. Milward (1977: 178); Deese (1981: 530).

35. Vernon (1983: 89); Kimura (1995: 555).

36. Painter (1986); Yergin (1991).

37. Navarro (1983); Kapstein (1990); Dulles and Dubofsky (1993).

38. Supple (1987: 536); Michel (1990: 271); Holter (1992); Gillingham (1991: 189); Kealey (1995); Gollan (1963); Levine (1958).

39. Giri (1958: 24); Sen (1994: 118); Kaur (1990: 85).

40. Pycroft and Munslow (1988: 158); Phimister (1994); Brown (1988: 54).

41. In this figure, the dimension of strike activity shown is worker participation in strikes, deflated by the size of each workforce. Worker participation rates are generally more indicative of the intensity of militancy than frequency rates. Graphs of numbers of strikes or days lost through strikes reveal even more starkly the strike-proneness of coal in comparison to oil. These patterns hold true for the post-1946 period as well.

42. The database also contains mentions from the *New York Times* for the period 1870–1905. Since these data are not strictly comparable to the post-1905 series, however, it is not included in the graph. See Appendix C for a more complete description of the data set.

43. See Edwards and Heery (1989) for a good discussion of the British example; Reid (1985) and Nowell (1994) on France; Navarro (1983) on the United States; Hein (1990) on Japan; and Gordon (1994) on Russia.

Chapter Five

1. On strategic energy planning in the United States, see Keohane (1984) and Painter (1986). For analyses of Soviet policies, see Lavigne (1983) and Reisinger (1992).

2. Moskoff (1973); Reisinger (1992).

3. On U.S. intervention in the Middle East, see Painter (1986); in Indonesia, see Russett (1985); in Latin America and elsewhere, see Knape (1987) and Frieden (1989).

4. For more on the American shift to more subtle forms of influence in the international oil system, see Venn (1986) and Bromley (1991).

5. For more on the role of oil in the postwar U.S. hegemonic order, see Keohane (1984) and Bromley (1991).

6. Knape (1987); Silverfarb (1996).

7. De Moraes (1993); Heiss (1994).

8. For details on the Egyptian crisis, see Lenczowski (1960: 321–36), Deese (1981: 554), Lipschutz (1989: 172), and Lustick (1997: 667).

9. Venn (1986); Silverfarb (1996).

10. Moran (1987) provides an excellent analysis of the oil order, which is informed by theories of hegemony. The information on multinational control over global oil reserves presented at the end of the paragraph comes from Bromley (1991: 97).

11. See Adelman (1972, 1995) and Beenstock and Vergottis (1993: 27–35) for more information on postwar investments in petroleum infrastructure. The information on private investments comes from Clark (1990: 121).

12. For additional information on the use of natural gas in ancient civilizations, see Temple (1986) and Riva and Atwater (2002).

13. Troxel (1937); Teece (1990); Dahl and Matson (1998).

14. Lyon (1990); Olien (2002).

15. Nash (1968); Olien (2002).

16. Clark (1990: 253); Melamid (1994).

17. Dienes (1971); Davison, Hurt, and Mabro (1988).

18. For historical information on the history of atomic science, see Romer (1982); for details of the weapons programs, see Rhodes (1988).

19. For analyses of this pronuclear military-industrial complex, consult Geiger (1992), Leslie (1993), and Winkler (1993).

20. Temples (1980: 242); Veiluva (1994: 126); Ecstein (1997: 35).

21. The Price-Anderson Act continues to protect the American civilian nuclear industry. Currently, the maximum pool of resources available to cover any accident or attack at a nuclear power plant is approximately $10 billion.

22. Lowen (1987: 479); Winkler (1993: 328); Nuclear Energy Institute (2003).

23. Clarfield and Wiecek (1984: 184).

24. On the strategic motivations behind civilian nuclear power development, see: Cowan (1990), Dauvergne (1993), Harrison (1996), and Takagi (1996).

25. Cochran, Norris, and Bukharin (1995); Bukharin (1997).

26. Mathieson (1980); Reisinger (1992); and Marples (1997).

27. Duffy (1978); Chang (1988).

28. For information on Indian nuclear developments, see Marwah (1977) and Noorani (1981). Goheen (1983) provides information on the Pakistani nuclear sector, while Broad, Sanger, and Bonner (2004) describe the proliferation efforts headed by A. Q. Khan.

29. Carter, Kanter, Perry, and Scowcroft (2003).

30. Data on energy consumption patterns immediately after the Second World War come from the United Nations (1952: 8); data from the 1970s come from Clark (1990: 115).

Chapter Six

1. See the following sources for more on the problematic impacts of oil-based development: Amuzegar (1982), Burke and Lubeck (1987), and Coronil (1997).

2. See Nore (1980) and Bergquist (1986) for analyses of the positional power enjoyed by oil workers in peripheral societies.

3. Lenczowski (1960); Lieuwen (1985); Karshenas (1990); Loyola-Diaz (1991); Heiss (1994).

4. See Bergesen and Schoenberg (1980) and Silver and Slater (1999) for discussions of these successive waves of decolonization.

5. Lieuwen (1985: 193); Alnasrawi (1991); Al-Azmeh (1995: 10).

6. Masters, Attanasi, and Root (1994); Campbell and Laherrere (1998).

7. Moran (1987); Yergin (1991: 531); Baddour (1997).

8. Consult Philip (1982: 84), Yergin (1991: 532), and Adelman (1995) for estimates on oil company profit rates.

9. Penrose (1987); Alnasrawi (1991).

10. Arrighi, Hopkins, and Wallerstein (1989); McLauchlan (1997).

11. Nye (1981); Ahrari and Noyes (1993).

12. Turner (1983: 133); Wall (1988: 859).

13. Jodice (1980); Kobrin (1993); Minor (1994).

14. Roncaglia (1983); Bina (1985); Alhajji and Huettner (2000).

15. See El Mallakh (1978) and Sirageldin (1988) for optimistic appraisals of the political and economic power of OPEC members in the wake of the price adjustments. Chalabi (1997), meanwhile, presents a survey of the diminishing political influence of OPEC states from the late 1980s onward.

16. Bobrow and Kudrle (1979); Ikenberry (1986).

17. At the time of its formation in 1974, the IEA had the following members: Austria, Belgium, Canada, Denmark, West Germany, Ireland, Japan, Luxembourg, the Netherlands, Spain, Sweden, Switzerland, the United Kingdom, and the United States.

18. Lieber (1979); Eguchi (1980); Renner (1984); Ikenberry (1986).

19. All IEA figures in this and the next paragraph report on inflation-adjusted dollars, with the year 2002 as the reference year. The data comes from the online IEA energy R&D database (IEA 2004).

20. See Anderson (1980), Coburn (1992), Organization for Economic Cooperation and Development (OECD) (1993), and Long (1997) for reviews of regulations addressing air pollution in OECD countries.

21. On the role of scientists in raising environmental concerns, see Miguel (1991) and Meyer et al. (1997: 623); on the link between public opinion and government environmental action, see Freudenburg and Rosa (1984), Farhar (1994), and Porter and Brown (1996); on factors behind corporate environmentalism, see Jones and Baldwin (1994: 13), Arora and Cason (1996), and Schmidheiny and Zorraquin (1996); and on radical environmentalism, see Wapner (1996).

22. Sandbach (1978); Reich (1983: 202); Rohrschneider (1991); O'Riordan and Clark (1995); Weidner (1995); McCully (1996); Kovarik (2001); Takao (2001: 292).

23. McCully (1996); Khagram (2003).

24. Navickis (1978); Plotkin (1980).

25. World Energy Council (1994); Mock, Tester, and Wright (1997).

26. Flavin (1981); Bodamer (1999); Brown (1999); Chiba (1996).

27. Perlin and Butti (2002); Kryza (2003).

28. Maycock (1984); U.S. Department of Energy (2002).

29. Hoffman, Wells, and Guiney (1998: 4); Perlin and Butti (2002).

30. Maycock (1984: 209); U.S. Department of Energy (2002).

31. See Berman and O'Connor (1996) for a good analysis of the rise and fall of this solar advocacy movement. Work by Lovins (1977) and other academics provided intellectual support for these grassroots efforts to spur the development of decentralized, smaller-scale energy systems.

32. The following sources provide useful descriptions of fuel cell systems: Flavin and Lenssen (1994), Difiglio (1997), Elliot (1997), and Rifkin (2002). Claims about the revolutionary potential of fuel cells have been made not only by environmentalists, but also by chief executive officers of oil corporations such as ARCO, Shell, and Texaco (Jewett 1999; Law 1999).

33. For reviews of the history of the fuel cell, see Koppel (1999) and Hoffmann (2002).

34. Tokumoto (1995); Moore (1997); Talalay (1997).

35. Dunn (2001); Hoffmann (2002).

36. Meyers and Schipper (1992); Lutzenhiser (1993).

37. Ghani (1998); Katouzian (2000).

38. Haykal (1982); Keddie (2003).

39. Painter (1986: 180); Moghadam (1988).

40. Abrahamian (1982); Keddie (2003).

41. Ramazani (1980); Parsa (1988).

42. Burke and Lubeck (1987); Moghadam (1988).

43. Ramazani (1980); Bayat (1998).

44. Parsa (1988); Keddie (2003).

45. Abrahamian (1980); Parsa (1988).

46. See Foran and Goodwin (1993) for a good summary of the preconditions that have preceded most revolutions in the modern period.

47. See Silver (2003) for more information on the World Labor Group database.

48. Parsa (1988: 68); Moaddel (1991).

49. Rubinstein (1981); Keddie (2003).

50. Riposa (1988); Patterson (1990); Morse (1992).

51. By 1990 the members of the IEA were Australia, Austria, Belgium, Canada, Denmark, West Germany, Greece, Ireland, Italy, Japan, Luxembourg, the Netherlands,

New Zealand, Portugal, Spain, Sweden, Switzerland, the United Kingdom, and the United States. All monetary information given is in inflation-adjusted, 2002 U.S. dollars.

52. Riposa (1988); Smith (1990).

53. Kamieniecki (1991); Rucht (1995).

54. Rosa and Dunlap (1994: 296); Dauvergne (1993: 580); de Boer and Catsburg (1988: 259).

55. The American study on nuclear plant costs was conducted by Komanoff and Roelofs (1992). All figures are in inflation-adjusted 1990 dollars. The British analysis is reported by Patterson (1990).

56. For reports on the documents that were declassified in 2003, see National Security Archive (2003), Marquis (2003), and Priest (2003).

57. See Bairoch (1993: 161), Maddison (1995), and Goldstein, Huang, and Akan (1997: 253–60) for analyses of the relationship between the oil price hikes and growing indebtedness in the developing world.

58. See Arrighi (1994; 2003) and Silver (2003) for more elaborate discussions of this monetarist counterrevolution.

59. Fearon (1988); Gill and Law (1989); Pastor (1989).

60. Luke (1983); Gately, Adelman, and Griffin (1986); Case (1995).

Chapter Seven

1. U.S. Energy Information Administration (2003).

2. See Goodstein (2004) for a review of the logic behind Hubbert's analysis and its relevance to contemporary oil production patterns.

3. Campbell and Laherrere (1998) provide the pessimistic estimate, while the U.S. Geological Survey (2000) provides the optimistic analysis.

4. Looney and Winterford (1995); Nitzan and Bichler (1995).

5. See Pew Research Center (2004) for survey data demonstrating rising hostility toward the US in the region.

6. Ogutcu (1998); Kinzer (1999).

7. On the expansion in coal, see Freese (2003). For a discussion of heavy oil projects, see Ogutcu (1998). For a review of fission and fusion developments, see Chang (2004).

8. For the United States, see U.S. Environmental Protection Agency (1997); for Mexico, see Eskeland (1992); for India, see Kumar (1997); for China, see Florig (1997); and for the economic estimates, see Leitmann (1999: 15–6).

9. World Meteorological Organization (2003).

10. Consult Marland et al. (1999) for data on greenhouse gas emissions.

11. Buchanan and Brenkley (1994); Tavoulareas and Charpentier (1995).

12. Pearce (2002).

13. See Quinn (2002), and the Database on Nuclear Smuggling, Theft, and Orphan Radiation Sources maintained at Stanford University, for information on this spread of nuclear materials.

14. See the Associated Press (2004) for a report on urban monitoring for radioactive materials.

15. Chiba (1996); Bull and Billman (2000: 240).

16. Schipper et al. (2001); Joskow (2002).

17. Perez-Pena (1996); Fouda (1998).

18. See Jones (2003) on the methane in lakes; see Rothstein (2003) on the methane off Japan's coast; and see Smil (2003) on global estimates of methane hydrate reserves.

19. Klinkenborg (2003).

20. See Koppel (1999), Hoffmann (2002), and Rifkin (2002) for very useful, if sometimes overly enthusiastic, discussions of the fuel cell's potential.

21. Tonn and Das (2002); Tromp et al. (2003).

22. See Jevons (1906) for the original articulation of the paradox, and Foster (2000) for a discussion of its contemporary relevance.

23. Data on the fuel cell survey are reported in Cropper (2003) and Cropper, Geiger, and Joilie (2003).

24. See Moore (1997), Hughes (1999), Siemens (1999), and Wald (1999) for overviews of investments in fuel cell systems made by established electrical manufacturing corporations.

25. For information on the Ballard/DaimlerChrysler/Ford partnership, see Law (1999) and Naus (1999); for the GM/Toyota partnership, see Brown (1999); and for fuel cell investments by other auto companies, see Ball (1999), Burt (1999), and Evarts (1999). The projection about mass production is offered by the European Hydrogen Association (2002).

26. Hakim (2003); J. D. Power & Associates (2003).

27. See the World Commission on Dams (2000) report for the survey results. McCully (1996) and Elliott (2003) provide critical analyses of large-scale dams.

28. In 2000 OECD member states included Australia, Austria, Belgium, Canada, the Czech Republic, Denmark, Finland, France, Germany, Hungary, Iceland, Ireland, Italy, Japan, Luxembourg, Mexico, the Netherlands, New Zealand, Norway, Poland, Portugal, South Korea, Slovakia, Spain, Sweden, Switzerland, Turkey, the United Kingdom, and the United States.

29. For more information on wind sectors, see McGowan and Connors (2000) and Hoge (1999). Details on solar sectors are provided by Schowengerdt (1999), Pearce (2002), and Geller (2003).

30. See reports from the World Energy Council (1990, 1994) on hydrogen's potential. Ogden, Williams, and Larson (2001: 6) provide the hydrogen/gasoline cost comparison information.

31. See Ogden (1999) and the National Academy of Sciences (2004) report for analyses of hydrogen production and distribution potentials.

32. See Nakicenovic, Grubler, and McDonald (1998: 57) for descriptions of other forecasts.

33. Reuters (2001); Llanos (2002); Bradsher (2003).

34. For an analysis of the successes that nongovernmental organizations have had in forcing the World Bank to reform its operations, see Rich (1994); for a detailed description of new World Bank energy policies, see World Bank (1999); and for European Union efforts to reform World Bank energy subsidies, see Kahn (2001).

35. Palcan Fuel Cells (2003); Geiger (2004).

36. Dailami and Leipziger (1997); Nakicenovic, Grubler, and McDonald (1998); Jensen and Ross (2000).

37. Van Natta and Butler (2003); McGeary (2004).

38. Peterson (2002); Cordesman (2003).
39. See AFL-CIO (1999) for an example of this union-led campaign.
40. Colley (1997); ICEM (2001).
41. Associated Press (2003); Cook (2004).
42. Gedicks (1995); Keck and Sikkink (1998); Forero (2003).
43. For an exciting example of how alternative energy systems can enhance the lives of people, see the example of Gaviotas, Colombia (Weisman 1998).
44. See the volume edited by Bornschier and Chase-Dunn (1999) for contrasting analyses of how such conflicts might come about.

Bibliography

Abrahamian, Ervand. 1980. "Structural Causes of the Iranian Revolution." *MERIP Reports* 87: 21–6.

———. 1982. *Iran between Two Revolutions*. Princeton, NJ: Princeton University Press.

Adelman, Morris. 1972. *The World Petroleum Market*. Baltimore, MD: Johns Hopkins University Press.

———. 1995. *The Genie out of the Bottle: World Oil Since 1970*. Cambridge, MA: MIT Press.

AFL-CIO. 1999. "US Energy Policy." *AFL-CIO Executive Council Actions (Press Release)*, Feb. 17, 1999.

Ahrari, M., and James Noyes. 1993. "Introduction: Background and Overview." In *The Persian Gulf after the Cold War*, edited by M. Ahrari and James Noyes, 1–15. Westport, CN: Praeger.

Al-Azmeh, Aziz. 1995. "Nationalism and the Arabs." *Arab Studies Quarterly* 17(1):1–18.

Aldcroft, D. 1964. "The Entrepreneur and the British Economy, 1870-1914." *Economic History Review* 17(1):113–34.

Alhajji, A., and David Huettner. 2000. "OPEC and World Crude Oil Markets from 1973 to 1994: Cartel, Oligopoly, or Competitive?" *Energy Journal* 21(3):31–60.

Aliyev, Natig. 1994. "The History of Oil in Azerbaijan." *Azerbaijan International* 2(2):1–10.

Alnasrawi, Abbas. 1991. *Arab Nationalism, Oil, and the Political Economy of Dependency*. New York, NY: Greenwood Press.

Amuzegar, J. 1982. "Oil and Wealth: A Very Mixed Blessing," *Foreign Affairs* 60(4):814–35.

Anderson, Douglas. 1980. "State Regulation of Electric Utilities." In *The Politics of Regulation*, edited by James Wilson, 3–40. New York, NY: Basic Books.

Arnot, D. 1866. "Coal and Smoke." *The Quarterly Review*, 515–35.

Arnot, R. 1975. *A History of the South Wales Miners' Federation, 1914–1926*. Cardiff, Australia: Cymric Federation Press.

Arora, Seema, and Timothy Cason. 1996. "Why Do Firms Volunteer to Exceed Environmental Regulations?" *Land Economics* 72(4):413–32.

Arrighi, Giovanni. 1994. *The Long Twentieth Century: Money, Power, and the Origins of Our Times*. London: Verso.

Arrighi, Giovanni, Terence Hopkins, and Immanuel Wallerstein. 1989. *Antisystemic Movements*. London: Verso.

Arrighi, Giovanni, and Beverly Silver. 1984. "Labor Movements and Capital Migration: The US and Western Europe in World-Historical Perspective." In *Labor in the Capitalist World-Economy*, edited by Charles Bergquist, 183–216. Beverly Hills, CA: Sage.

———. 1999. *Chaos and Governance in the Modern World System*. Minneapolis: University of Minnesota Press.

Associated Press. 2003. "China Orders Stepped-Up Emissions Controls at Coal-Fired Power Plants." *Associated Press*, Oct. 9.

———. 2004. "Nuclear Experts Search for Dirty Bombs." *New York Times*, Jan. 7.

Baddour, J. 1997. "The International Petroleum Industry." *Energy Policy* 25(2): 143–57.

Bairoch, Paul. 1993. *Economics and World History: Myths and Paradoxes*. New York: Harvester Wheatsheaf.

Ball, Jeffrey. 1999. "Auto Makers Are Racing to Market 'Green' Cars Powered by Fuel Cells." *Wall Street Journal*, Mar. 15, A1–A8.

Bamberg, J. 1994. *The History of the British Petroleum Company, Volume 2: The Anglo-Iranian Years, 1928–1954*. Cambridge: Cambridge University Press.

Banks, Ferdinand. 1985. *The Political Economy of Coal*. Lexington, MA: Heath and Company.

Barker, T. 1985. "The International History of Motor Transport." *Journal of Contemporary History* 20(1):3–19.

Barrie, Alexander. 1961. *War Underground*. New York: Ballantine Press.

Bayat, Asef. 1998. "Revolution without Movement, Movement without Revolution: Comparing Islamic Activism in Iran and Egypt." *Comparative Studies in Society and History* 40(1):136–69.

Beenstock, Michael, and Andreas Vergottis. 1993. *Econometric Modeling of World Shipping*. London: Chapman and Hall.

Bergesen, Albert, and Ronald Schoenberg. 1980. "Long Waves of Colonial Expansion and Contraction, 1415–1969." In *Studies of the Modern World System*, edited by Albert Bergesen, 231–77. New York: Academic Press.

Bergquist, Charles. 1986. *Labor in Latin America: Comparative Essays on Chile, Argentina, Venezuela, and Colombia*. Stanford, CA: Stanford University Press.

Berman, Daniel, and John O'Connor. 1996. *Who Owns the Sun? People, Politics, and the Struggle for a Solar Economy*. White River Junction, VT: Chelsea Green Publishing Company.

Bina, Cyrus. 1985. *The Economics of the Oil Crisis: Theories of Oil Crisis, Oil Rent, and Internationalization of Capital in the Oil Industry*. London: Merlin Press.

Blatz, Perry. 1994. *Democratic Miners: Work and Labor Relations in the Anthracite Coal Industry, 1875–1925*. Albany, NY: State University of New York Press.

Bloomfield, Arthur. 1968. *Patterns of Fluctuation in International Investment before 1914*. Princeton, NJ: Princeton University Press.

Bobrow, Davis, and Robert Kudrle. 1979. "Energy R&D: In Tepid Pursuit of Common Goals." *International Organization* 33(2):149–75.

Bodamer, David. 1999. "Catch the Wind." *Civil Engineering* 69(7):50–4.

Bornschier, Volker, and Christopher Chase Dunn. 1999. *The Future of Global Conflict*. Thousand Oaks, CA: Sage.

Boswell, Terry, and Christopher Chase-Dunn. 2000. *The Spiral of Capitalism and Socialism: Toward Global Democracy*. Boulder, CO: Lynne Rienner Publishers, Inc.

Bowen, Ralph. 1950. "Rise of Modern Industry: The Roles of Government and Private Enterprise in German Industrial Growth, 1870–1914." *Journal of Economic History* 10:68–81.

Bowman, John. 1989. *Capitalist Collective Action: Competition, Cooperation, and Conflict in the Coal Industry*. Cambridge: Cambridge University Press.

Boyns, Trevor, and Judith Wale. 1996. "The Development of Management Information Systems in the British Coal Industry, 1880–1947." *Business History* 38(2):55–77.

Bradsher, Keith. 2003. "China Set to Act on Fuel Economy." *New York Times*, Nov. 18.

Braudel, Fernand. 1992. *The Structures of Everyday Life: Civilization and Capitalism, Volume I*. Berkeley, CA: University of California Press.

Brenner, Robert. 2002. *The Boom and the Bubble: The US in the World Economy*. New York: Verso.

British Petroleum Company. 2003. *BP Statistical Review of World Energy*. London: British Petroleum.

Brittain, James. 1974. "The International Diffusion of Electrical Power Technology, 1870–1920." *Journal of Economic History* 34(1):108–21.

Broad, William, David Sanger, and Raymond Bonner. 2004. "A Tale of Nuclear Proliferation." *New York Times*, Feb. 12.

Broadbridge, Seymour. 1989. "Aspects of Economic and Social Policy in Japan, 1868–1945." In *The Cambridge Economic History of Europe, Volume VIII*, edited by Peter Mathias and Sidney Pollard, 1106–45. Cambridge: Cambridge University Press.

Brody, David. 1990. "Labour Relations in American Coal Mining." In *Workers, Owners and Politics in Coal Mining*, edited by Gerald Feldman and Klaus Tenfelde, 74–117. New York: St. Martin's Press.

Bromley, Simon. 1991. *American Hegemony and World Oil*. University Park, PA: Pennsylvania State University Press.

Brown, Carolyn. 1988. "The Dialectics of Colonial Labour Control: Class Struggles in the Nigerian Coal Industry, 1914–1949." *Journal of Asian and African Studies* 23(1):32–55.

Brown, Jonathan. 1993. *Oil and Revolution in Mexico*. Berkeley, CA: University of California Press.

Brown, Richard. 1991. *Society and Economy in Modern Britain 1700–1850*. London: Routledge.

Brown, Warren. 1999. "GM, Toyota Team Up to Make Electric Cars." *Washington Post*, April 20.

Buchanan, D., and D. Brenkley. 1994. "Green Coal Mining." In *Mining and Its Environmental Impact*, edited by R. Hester and R. Harrison. Cambridge, MA: Royal Society of Chemistry.

Buchanan, Robert. 2003. "The History of Technology." *Encyclopedia Britannica*.

Bukharin, Oleg. 1997. "The Future of Russia's Plutonium Cities." *International Security* 21(4):126–58.

Bull, Stanley, and Lynn Billman. 2000. "Renewable Energy: Ready to Meet its Promise?" *Washington Quarterly* 23(1):229–44.

Bunker, Stephen, and Denis O'Hearn. 1993. "Strategies of Economic Ascendants for Access to Raw Materials: A Comparison of the United States and Japan." In *Pacific-Asia and the Future of the World-System*, edited by Ravi Palat. Westport, CN: Greenwood Press.

Bunker, Stephen, and Paul Ciccantell. 1999. "Economic Ascent and the Global Environment: World-Systems Theory and the New Historical Materialism." In *Ecology and the World-System*, edited by Walter Goldfrank, David Goodman, and Andrew Szasz, 107–22. Westport, CN: Greenwood Press.

Burke, Edmund, and Paul Lubeck. 1987. "Explaining Social Movements in Two Oil-Exporting States: Divergent Outcomes in Nigeria and Iran." *Comparative Studies in Society and History* 29(4):643–65.

Burt, Roger. 1991. "The International Diffusion of Technology in the Early Modern Period: The Case of the British Non-Ferrous Mining Industry." *Economic History Review* 44(2):249–71.

Burt, Tim. 1999. "Car Makers in Fuel Cells Initiative." *Financial Times (London)*, Sept. 20.

Cain, P., and A. Hopkins. 1980. "The Political Economy of British Expansion Overseas, 1750–1914." *Economic History Review* 33(4):463–90.

Cairncross, A. 1953. *Home and Foreign Investment 1870–1913*. Cambridge: Cambridge University Press.

Cameron, Rondo. 1997. *A Concise Economic History of the World*. Oxford: Oxford University Press.

Campbell, Colin. 1991. *The Golden Century of Oil, 1950–2050: The Depletion of a Resource*. Dordrecht: Kluwer Academic Publishers.

Campbell, Colin, and Jean Laherrere. 1998. "The End of Cheap Oil." *Scientific American* March: 78–3.

Cantor, Geoffrey, David Gooding, and Frank James. 1996. *Michael Faraday*. Atlantic Highlands, NJ: Humanities Press.

Carter, Ashton, Arnold Kanter, William Perry, and Brent Scowcroft. 2003. "Good Nukes, Bad Nukes." *New York Times*, Dec. 22.

Carver, T. 1924. *The Economy of Human Energy*. New York: MacMillan.

Case, J. 1995. "Oil Pricing." In *The New Global Oil Market: Understanding Energy Issues in the World Economy*, edited by Siamack Shojai. Westport, CN: Praeger.

Center for Strategic and International Studies. 2001. *The Geopolitics of Energy into the 21st Century: The Report of the CSIS Strategic Energy Initiative*. Washington, DC: CSIS.

Chalabi, Fadhil. 1997. "OPEC: An Obituary." *Foreign Policy* 109:126–40.

Chandler, Alfred. 1962. *Strategy and Structure: Chapters in the History of the Industrial Enterprise*. Cambridge, MA: MIT Press.

———. 1972. "Anthracite Coal and the Beginnings of the Industrial Revolution in the United States." *Business History Review* 46(2):141–81.

———. 1990. *Scale and Scope: The Dynamics of Industrial Capitalism*. Cambridge, MA: Belknap Press of Harvard University.

Chang, Gordon. 1988. "JFK, China, and the Bomb." *Journal of American History* 74(4):1287–1310.

Chang, Kenneth. 2004. "Experts Say New Desktop Fusion Claims More Credible." *New York Times*, March 3.

Chase-Dunn, Christopher. 1989. *Global Formation: Structures of the World-Economy*. Cambridge, MA: Basil Blackwell.

Chase-Dunn, Christopher, and Bruce Podobnik. 1999. "The Next World War: World-System Cycles and Trends." In *The Future of Global Conflict*, edited by Volker Bornschier and Christopher Chase-Dunn. London: Sage.

Chew, Sing. 1995. "Environmental Transformations: Accumulation, Ecological Crisis, and Social Movements." In *A New World Order?* Edited by David Smith and Jozsef Borocz, 201–15. Westport, CN: Praeger.

Chiba, Mitsugi. 1996. "New Sunshine Program." *Journal of Energy Engineering* 122(3):93–102.

Choucri, Nazli, and Robert North. 1975. *Nations in Conflict: National Growth and International Violence*. San Francisco: W. H. Freeman and Company.

Church, Phillip. 1971. "Labor Relations in Mineral and Petroleum Resource Development." In *Foreign Investment in the Petroleum and Mineral Industries*, edited by Raymond Mikesell, 81–98. Baltimore, MD: Johns Hopkins Press.

Church, Roy. 1986. *The History of the British Coal Industry, Volume 3*. Oxford: Clarendon Press.

Cipolla, Carlo. 1976. *Fontana Economic History of Europe*. London: Collins/ Fontana.

Clarfield, Gerard, and William Wiecek. 1984. *Nuclear America: Military and Civilian Nuclear Power in the United States, 1940–1980*. New York: Harper & Row.

Clark, John. 1990. *The Political Economy of World Energy*. Chapel Hill, NC: University of North Carolina Press.

Clemens, Paul. 1976. "The Rise of Liverpool, 1665–1750." *Economic History Review* 29(2):211–25.

Coburn, Leonard. 1992. "The Clean Air Act Amendments of 1992." In *International Issues in Energy Policy, Development and Economics*, edited by James Dorian and Fereidun Fesharaki. Boulder, CO: Westview Press.

Cochran, Thomas, Robert Norris, and Oleg Bukharin. 1995. *Making the Russian Bomb: From Stalin to Yeltsin*. Boulder, CO: Westview Press.

Cohn, Samuel. 1993. *When Strikes Make Sense—And Why*. New York: Plenum Press.

Colley, Peter. 1997. *Reforming Energy: Sustainable Futures and Global Labour*. London: Pluto Press.

Cook, Terry. 2004. "Drive for Coal Produces More Deaths in China's Mines." *China Labour Bulletin (Hong Kong)* March 16:1–2.

Cooling, Benjamin. 1979. *Gray Steel and Blue Water Navy: The Formative Years of America's Military-Industrial Complex, 1881–1917*. Hamden, CN: Archon Books.

Cordesman, Anthony. 2003. *Saudi Arabia Enters the Twenty-First Century*. Westport, CN: Praeger.

Coronil, Fernando. 1997. *The Magical State: Nature, Money and Modernity in Venezuela*. Chicago, IL: University of Chicago Press.

Cottrell, Fred. 1955. *Energy and Society*. New York: McGraw Hill.

Cowan, Robin. 1990. "Nuclear Power Reactors: A Study in Technological Lock-In." *Journal of Economic History* 50(3):541–67.

Criqui, Patrick. 1994. "Energy Crises and Economic Crisis: A Long-Period Perspective." *Energy Studies Review* 6(1):34–46.

Cromar, Peter. 1977. "The Coal Industry on Tyneside." *Economic Geography* 53(1):79–94.

Cropper, Mark. 2003. "Fuel Cell Market Survey: Buses," *Fuel Cell Today*, November 26, pp. 1–12.

Cropper, Mark, Stefan Geiger, and David Joilie. 2003. "Fuel Cell Systems: A Survey of Worldwide Activity." *Fuel Cell Today* Nov. 5:1–12.

Crouzet, Francois. 1990. *Britain Ascendant: Comparative Studies in Franco-British Economic History*. Cambridge: Cambridge University Press.

Cummins, C. 1989. *Internal Fire*. Warrendale, PA: Society of Automotive Engineers.

Dahl, Carol, and Thomas Matson. 1998. "Evolution of the US Natural Gas Industry." *Land Economics* 74(3):390–408.

Dailami, Mansoor, and Danny Leipziger. 1997. *Infrastructure Project Finance and Capital Flows: A New Perspective*. Washington, DC: World Bank.

Darmstadter, Joel, Perry Teitelbaum, and Jaroslav Polach. 1971. *Energy in the World Economy: A Statistical Review of Trends in Output, Trade, and Consumption since 1925*. Baltimore, MD: Johns Hopkins Press.

Dauvergne, Peter. 1993. "Nuclear Power Development in Japan." *Asian Survey* 33(6):576–91.

Davidson, Ray. 1988. *Challenging the Giants: A History of the Oil, Chemical and Atomic Workers International Union*. Denver, CO: Oil, Chemical and Atomic Workers International Union.

Davison, Ann, Chris Hurt, and Robert Mabro. 1988. *Natural Gas: Governments and Oil Companies in the Third World*. Oxford: Oxford University Press.

De Boer, Connie, and Ineke Catsburg. 1988. "The Impact of Nuclear Accidents on Attitudes toward Nuclear Energy." *Public Opinion Quarterly* 52(2):254–61.

De Moraes, Moyara. 1993. "Operation 'Ajax' Revisited: Iran, 1953." *Middle Eastern Studies* 29(3):467–86.

Debeir, Jean Claude, Jean Paul Deleage, and Daniel Hemery. 1991. *In the Servitude of Power: Energy and Civilization Through the Ages*. London: Zed Books.

Deese, David. 1981. "Oil, War and Grand Strategy." *Orbis* 25(3):525–55.

Dennis, Norman, Fernando Henriques, and Clifford Slaughter. 1956. *Coal Is Our Life: An Analysis of a Yorkshire Mining Community*. London: Tavistock Publications.

DeNovo, John. 1955. "Petroleum and the United States Navy before World War I." *Mississippi Valley Historical Review* 41(4):641–56.

———. 1956. "The Movement for an Aggressive American Oil Policy Abroad, 1918–1920." *American Historical Review* 61(4):854–76.

Devine, T. 1976. "The Colonial Trade and Industrial Investment in Scotland." *Economic History Review* 29(1):1–13.

Devine, Warren. 1983. "From Shafts to Wires." *Journal of Economic History* 43(2): 347–72.

Dickson, Peter. 1967. *The Financial Revolution in England*. London: MacMillan.

Dienes, Leslie. 1971. "Issues in Soviet Energy Policy." *Soviet Studies* 23(1):26–58.

Difiglio, Carmen. 1997. "Using Advanced Technologies to Reduce Motor Vehicle Greenhouse Gas Emissions." *Energy Policy* 25(14):1173–78.

Dix, Keith. 1988. *What's a Coal Miner to Do? The Mechanization of Coal Mining*. Pittsburgh, PA: University of Pittsburgh Press.

Donnelly, Thomas. 2000. *Rebuilding America's Defenses*. Washington, DC: Project for the New American Century.

Duffy, Gloria. 1978. "Soviet Nuclear Exports." *International Security* 3(1):83–111.

Dulles, Foster, and Melvyn Dubofsky. 1993. *Labor in America: A History*. Wheeling, IL: Harlan Davidson, Inc.

Dumett, R. 1975. "Joseph Chamberlain, Imperial Finance and Railway Policy in British West Africa." *English Historical Review* 90(355):287–321.

Dunn, Seth. 2001. *Hydrogen Futures: Toward a Sustainable Energy System*. Washington: Worldwatch Institute.

Dunning, John, and Howard Archer. 1993. "The Eclectic Paradigm and the Growth of UK Multinational Enterprise 1870–1983." In *Transnational Corporations: A Historical Perspective*, edited by Geoffrey Jones. London: Routledge.

Eavenson, Howard. 1939. *Coal through the Ages*. New York: American Institute of Mining and Metallurgical Engineers.

Eckstein, Rick. 1997. *Nuclear Power and Social Power*. Philadelphia, PA: Temple University Press.

Edwards, Christine, and Edmund Heery. 1989. *Management Control and Union Power: A Study of Labour Relations in Coal-Mining*. Oxford: Clarendon Press.

Edwards, Paul. 1981. *Strikes in the United States, 1881–1974*. New York: St. Martin's Press.

Eguchi, Yujiro. 1980. "Japanese Energy Policy." *International Affairs* 56(2):263–79.

El Mallakh, Ragaei. 1978. "Prospects for Economic Growth and Regional Coop-eration." In *The Middle East in the Coming Decade*, edited by John Waterbury and Ragaei El Mallakh, 149–206. New York: McGraw-Hill.

Elliot, David. 1997. *Energy, Society and Environment*. London: Routledge.

———. 2003. "Sustainable Energy: Choices, Problems and Opportunities." In *Sustainability and Environmental Impact of Renewable Energy Resources*, edited by R. Hester and R. Harrison, 19–48. Cambridge, MA: Royal Society of Chemistry.

Ellis, Richard. 1991. *Men and Whales*. New York: Knopf.

Emmons, William. 1993. "Franklin D. Roosevelt, Electric Utilities, and the Power of Competition." *Journal of Economic History* 53(4):880–907.

Encyclopedia of Social Sciences. 1911. *Motor Vehicles*. London: Encyclopedia Publishers, Ltd.

Eskeland, Gunnar. 1992. *The Objective: Reduce Pollution at Low Cost*. Washington, DC: World Bank Country Economics Department.

Etemad, Bouda, and Jean Luciani. 1991. *World Energy Production 1800–1985*. Geneva: Libraire DROZ.

European Hydrogen Association. 2002. "Hydrogen Economy to Fuel the 21st Century." *European Hydrogen Association*, Aug. 29.

Evarts, Eric. 1999. "The Refueling of America." *Christian Science Monitor*, April 22:13.

Fairbanks, Charles. 1991. "The Origins of the Dreadnought Revolution." *International History Review* 13(2):246–72.

Farhar, Barbara. 1994. "The Polls-Poll Trends: Public Opinion about Energy." *Public Opinion Quarterly* 58(4):603–32.

Fearon, James. 1988. "International Financial Institutions and Economic Policy Reform in Sub-Saharan Africa." *Journal of Modern Africa Studies* 26(1):113–37.

Fearon, Peter. 1985. "The Growth of Aviation in Britain." *Journal of Contemporary History* 20(1):21–40.

Feldman, Gerald. 1966. *Army, Industry and Labor in Germany, 1914–1918*. Princeton, NJ: Princeton University Press.

Feldman, Gerald, and Klaus Tenfelde. 1990. *Workers, Owners and Politics in Coal Mining*. New York: St. Martin's Press.

Ferguson, Eugene. 1967. "The Steam Engine before 1830." In *Technology in Western Civilization, Volume I*, edited by Melvin Kranzberg and Carroll Pursell. New York: Oxford University Press.

Ferrier, Ronald. 1982. *The History of the British Petroleum Company, Volume 1: The Developing Years, 1901–1932*. Cambridge: Cambridge University Press.

Fine, Ben. 1990. *The Coal Question: Political Economy and Industrial Change from the Nineteenth Century to the Present Day*. New York: Routlege.

Flavin, Christopher. 1981. "A Renaissance for Wind Power." *Environment* 23(8):31–42.

Flavin, Christopher, and Nicholas Lenssen. 1994. *Power Surge: Guide to the Coming Energy Revolution*. New York: W.W. Norton and Company.

Flinn, Michael, and V. Fogleman. 1984. *The History of the British Coal Industry, Volume 2*. Oxford: Clarendon Press.

Florig, H. 1997. "China's Air Pollution Risks." *Environmental Science & Technology* 31(6):274–79.

Fohlen, Claude. 1978. "Entrepreneurship and Management in France in the Nineteenth Century." In *The Cambridge Economic History of Europe*, edited by Peter Mathias and M. Postan, 347–81. Cambridge: Cambridge University Press.

Foran, John, and Jeff Goodwin. 1993. "Revolutionary Outcomes in Iran and Nicaragua." *Theory and Society* 22(2):209–47.

Forbes, R. 1958. *Studies in Early Petroleum History*. Leiden: E. J. Brill.

Foreman-Peck, James. 1982. "The American Challenge of the Twenties: Multinationals and the European Motor Industry." *Journal of Economic History* 42(4):865–81.

Forero, Juan. 2003. "Seeking Balance: Growth vs Culture in Amazon." *New York Times*, Dec. 10.

Foster, John Bellamy. 1999. "Marx's Theory of Metabolic Rift." *American Journal of Sociology* 105(2):366–405.

———. 2000. "Capitalism's Environmental Crisis—Is Technology the Answer?" *Monthly Review* 52(7):1–13.

Fouda, Safaa. 1998. "Liquid Fuels from Natural Gas." *Scientific American*, March: 92–5.

Fouquet, Roger, and Peter Pearson. 1998. "A Thousand Years of Energy Use in the United Kingdom." *Energy Journal* 19(4):1–41.

Freese, Barbara. 2003. *Coal: A Human History*. Cambridge, MA: Perseus.

Fremdling, Rainer. 1996. "Anglo-German Rivalry in Coal Markets in France, the Netherlands and Germany 1850–1913." *Journal of European Economic History* 25(3):599–645.

Freudenburg, W., and E. Rosa. 1984. *Public Reactions to Nuclear Power: Are There Critical Masses?* Boulder, CO: Westview Press.

Frieden, Jeffry. 1989. "The United States in the International Economy." *Comparative Studies in Society and History* 31(1):55–80.

Gately, Dermot, Morris Adelman, and James Griffin. 1986. "Lessons from the 1986 Oil Price Collapse." *Brookings Papers on Economic Activity* 2:237–84.

Geary, Dick. 1981. *European Labour Protest, 1848–1939*. New York: St. Martin's Press.

———. 1989. "Socialism and the German Labour Movement before 1914." In *Labour and Socialist Movements in Europe before 1914*, edited by Dick Geary, 101–27. New York: St. Martin's Press.

Geary, Roger. 1985. *Policing Industrial Disputes: 1893 to 1985*. New York: Cambridge University Press.

Gedicks, Al. 1995. "International Native Resistance to the New Resource Wars." In: *Ecological Resistance Movements*, edited by Bron Taylor, 89–108. Albany, NY: State University of New York Press.

Geiger, Roger. 1992. "Science, Universities, and National Defense, 1945–1970." *Osiris* 7:26–48.

Geiger, Stefan. 2004. "Hyundai presents new fuel cell SUV." *Fuel Cell Today*, April 7.

Geller, Howard. 2003. *Energy Revolution: Policies for a Sustainable Future.* Washington, DC: Island Press.

Ghani, Cyrus. 1998. *Iran and the Rise of Reza Shah.* London: I.B. Tauris.

Gibb, George, and Evelyn Knowlton. 1956. *History of Standard Oil Company.* New York: Harper.

Giebelhaus, August. 1982. "Petroleum Refining and Transportation." In *Energy and Transport: Historical Perspectives on Policy Issues,* edited by George Daniels and Mark Rose. Beverly Hills, CA: Sage Publications.

Gilbert, David. 1992. *Class, Community, and Collective Action: Social Change in British Coalfields.* Oxford: Clarendon Press.

Gill, Stephen, and David Law. 1989. "Global Hegemony and the Structural Power of Capital." *International Studies Quarterly* 33(4):475–99.

Gillingham, John. 1991. *Coal, Steel, and the Rebirth of Europe, 1945–1955.* Cambridge: Cambridge University Press.

Giri, V. 1958. *Labour Problems in Indian Industry.* Bombay: Asia Publishing House.

Goheen, Robert. 1983. "Problems of Proliferation: US Policy and the Third World." *World Politics* 35(2):194–215.

Goldstein, Joshua. 1988. *Long Cycles: Prosperity and War in the Modern Age.* New Haven, CN: Yale University Press.

Goldstein, Joshua, X. Huang, and Burcu Akan. 1997. "Energy in the World Economy, 1950–1992." *International Studies Quarterly* 41:241–66.

Gollan, Robin. 1963. *The Coalminers of New South Wales.* Melbourne: Melbourne University Press.

Goodrich, Carter. 1977. *The Miner's Freedom.* New York: Arno Press.

Goodstein, David. 2004. *Out of Gas.* New York: W. W. Norton & Company.

Gordon, David. 1980. "Stages of Accumulation and Long Economic Cycles." In *Processes of the World-System,* edited by Terence Hopkins and Immanuel Wallerstein, 9–45. Beverly Hills, CA: Sage.

Gordon, Leonid. 1994. "Russian Workers and Democracy." *International Journal of Sociology* 23(4):3–96.

Great Britain Central Statistical Office. 1996. *Annual Statistical Bulletin.* London.

Greenberg, Dolores. 1990. "Energy, Power, and Perceptions of Social Change in the Early Nineteenth Century." *American Historical Review* 95(3): 693–714.

Grieve, W., and Bernard Newman. 1936. *Tunnellers.* London: H. Jenkins, Ltd.

Griffiths, Denis. 1997. *Steam at Sea: Two Centuries of Steam-Powered Ships.* London: Conway Maritime Press.

Gross, Charles. 1990. "George Owen Squier and the Origins of American Military Aviation." *Journal of Military History* 54(3):281–306.

Hakim, Danny. 2003. "Hybrid Cars are Catching On." *New York Times,* Jan. 28.

Hallwood, Paul, and Stuart Sinclair. 1982. "OPEC's Developing Relationships with the Third World." *International Affairs* 58(2):271–86.

Hansen, James, Makiko Sato, Reto Ruedy, Andrew Lacis, and Valdar Oinas. 2000. "Global Warming in the Twenty-First Century: An Alternative Scenario." *Proceedings of the National Academy of Sciences* 97(18):9875–80.

Hard, M., and A. Jamison. 1997. "Alternative Cars: The Contrasting Stories of Steam and Diesel Automotive Engines." *Technology in Society* 19(2): 145–60.

Harding, G. 1946. "American Coal Production and Use." *Economic Geography* 22(1):46–53.

Harre, Rom. 1984. *Great Scientific Experiments*. Oxford: Oxford University Press.

Harrison, Selig. 1996. "Japan and Nuclear Weapons." In *Japan's Nuclear Future: The Plutonium Debate and East Asian Security*, edited by Selig Harrison. Washington, DC: Carnegie Endowment for International Peace.

Hartwell, Robert. 1960. "Markets, Technology, and the Structure of Enterprise in the Development of the Eleventh-Century Chinese Iron and Steel Industry." *Journal of Economic History* 26(1):29–58.

Harvey, David. 1982. *The Limits to Capital*. Oxford: Basil Blackwell.

———. 2003. *The New Imperialism*. Oxford: Oxford University Press.

Haykal, Muhammad. 1982. *Iran, the Untold Story*. New York: Pantheon Books.

Headrick, Daniel. 1981. *The Tools of Empire*. New York: Oxford University Press.

———. 1988. *The Tentacles of Progress*. New York: Oxford University Press.

Heaton, Herbert. 1967. "The Spread of the Industrial Revolution." In *Technology in Western Civilization, Volume I*, edited by Melvin Kranzberg and Carroll Pursell. New York: Oxford University Press.

Hein, Laura. 1990. *Fueling Growth*. Cambridge, MA: Harvard University Press.

Heinze, G., and Heinrich Kill. 1988. "The Development of the German Railroad System." In *The Development of Large Technical Systems*, edited by Renate Mayntz and Thomas Hughes. Boulder, CO: Westview Press.

Heiss, Mary. 1994. "The United States, Great Britain, and the Creation of the Iranian Oil Consortium, 1953–1954." *International History Review* 16(3): 511–35.

Henderson, W. 1948. "German Economic Penetration in the Middle East, 1870–1914." *Economic History Review* 18(1/2):54–64.

Henderson, William. 1975. *The Rise of German Industrial Power, 1834–1914*. Berkeley, CA: University of California Press.

Henning, Graydon, and Keith Trace. 1975. "Britain and the Motorship." *Journal of Economic History* 35(2):353–85.

Henriques, Robert. 1960. *Marcus Samuel: Founder of the Shell Transport and Trading Company*. London: Barrie and Rockliff.

Hentenryk, Ginette, and Jean Puissant. 1990. "Industrial Relations in the Belgian Coal Industry." In *Workers, Owners and Politics in Coal Mining*, edited by Gerald Feldman and Klaus Tenfelde, 203–70. New York: St. Martin's Press.

Heron, Craig. 1989. *The Canadian Labour Movement: A Short History*. Toronto: James Lorimer & Company.

Herwig, Holger. 1991. "The German Reaction to the Dreadnought Revolution." *International History Review* 13(2):273–83.

Hickey, S. 1985. *Workers in Imperial Germany*. New York: Oxford University Press.

Hilden, Patricia. 1993. *Women, Work, and Politics: Belgium, 1830–1914*. Oxford: Clarendon Press.

Hirsh, Richard. 1990. *Technology and Transformation in the American Electric Utility Industry*. Cambridge: Cambridge University Press.

Hobsbawm, Eric. 1962. *The Age of Revolution, 1789–1848*. New York: New American Library.

———. 1964. *Labouring Men: Studies in the History of Labour*. New York: Basic Books.

Hoffman, John, John Wells, and William Guiney. 1998. "Transforming the Market for Solar Water Heaters." *Renewable Energy Policy Project Research Report* 4(August 1998):1–20.

Hoffmann, Peter. 2002. *Tomorrow's Energy*. Cambridge, MA: MIT Press.

Hoge, Warren. 1999. "Denmark Moves Ahead in Wind Power." *New York Times*, Oct. 9.

Holter, Darryl. 1992. *The Battle for Coal*. Dekalb, IL: Northern Illinois University Press.

Hounshell, David. 1984. *From the American System to Mass Production 1800–1932*. Baltimore, MD: Johns Hopkins University Press.

Hourani, Albert. 1991. *A History of the Arab Peoples*. Cambridge, MA: Harvard University Press.

Howarth, Stephen. 1997. *A Century in Oil: The Shell Transport and Trading Company, 1897–1997*. London: Weidenfeld & Nicolson.

Howe, Christopher. 1996. *The Origins of Japanese Trade Supremacy*. London: Hurst & Company.

Hubbert, Marion. 1962. *Energy Resources*. Washington, DC: National Academy of Sciences.

Hughes, Claire. 1999. "Start-Up Hopes Investors, Public Plug Into Fuel-Cell Developments." *Plain Dealer*, Sept. 5:2E.

Hughes, Thomas. 1983. *Networks of Power: Electrification in Western Society, 1880–1930*. Baltimore, MD: Johns Hopkins University Press.

Hugill, Peter. 1993. *World Trade since 1431: Geography, Technology, and Capitalism*. Baltimore, MD: Johns Hopkins University Press.

Hunter, Louis. 1985. *A History of Industrial Power in the United States, 1780–1930: Volume Two: Steam Power*. Charlottesville, NC: University Press of Virginia.

Hunter, Louis, and Lynwood Bryant. 1991. *A History of Industrial Power in the United States, 1780–1930, Volume Three*. Cambridge, MA: MIT Press.

Hymer, Stephen. 1979. *The Multinational Corporation: A Radical Approach*. Cambridge: Cambridge University Press.

International Federation of Chemical, Energy, Mine and General Workers' Unions. 2001. *Labour and Climate Change*. Brussels: ICEM.

Ikenberry, John. 1986. "The Irony of State Strength: Comparative Responses to the Oil Shocks in the 1970s." *International Organization* 40(1):105–37.

———. 1988. *Reasons of State: Oil Politics and the Capacities of American Government*. Ithaca, NY: Cornell University Press.

Intergovernmental Panel on Climage Change. 2001. *Third Assessment Report of Working Group 1*. Shanghai: UN IPCC.

International Energy Agency. 1998. *Energy Prices and Taxes*. Paris: IEA/OECD.

———. 2003. *Renewables Information*. Paris: IEA.

———. 2004. *IEA Energy Technology RD&D Statistics*. Paris: IEA.

Isser, Steve. 1996. *The Economics and Politics of the United States Oil Industry, 1920–1990*. New York: Garland Publishing.

J. D. Power and Associates. 2003. "More than One-Half of the Hybrid Market May Be Trucks by 2008." *J. D. Power and Associates Reports*, Oct. 27.

Jamieson, Stuart. 1968. *Times of Trouble: Labour Unrest and Industrial Conflict in Canada, 1900–66*. Ottawa, Canada: Privy Council Office Task Force on Labour Relations.

Jensen, Marc, and Marc Ross. 2000. "The Ultimate Challenge: Developing an Infrastructure for Fuel Cell Vehicles." *Environment* 42(7):10–22.

Jensen, W. 1968. "The Importance of Energy in the First and Second World Wars." *Historical Journal* 11(3):538–54.

Jeremy, David. 1977. "Damming the Flood: British Government Efforts to Check the Outflow of Technicians and Machinery." *Business History Review* 51(1):1–34.

Jevons, William. 1865 (1906). *The Coal Question*. London: MacMillan.

Jewett, Dale. 1999. "ARCO Sees Future in Fuel Cells." *Detroit News*, April 27:B3.

Jodice, David. 1980. "Sources of Change in Third World Regimes for Foreign Direct Investment, 1968–1976." *International Organization* 34(2):177–206.

Johnson, Arthur. 1967. "Expansion of the Petroleum and Chemical Industries, 1880–1900." In *Technology in Western Civilization, Volume I*, edited by Melvin Kranzberg and Carroll Pursell. New York: Oxford University Press.

Jones, Geoffrey. 1981. *The State and the Emergence of the British Oil Industry*. London: MacMillan.

Jones, L., and John Baldwin. 1994. *Corporate Environmental Policy and Government Regulation*. London: JAI Press.

Jones, Nicola. 2003. "Lake Kivu Stores Enough Energy to Power All of Rwanda." *New Scientist* 177(March 1):17–8.

Joskow, Paul. 2002. "United States Energy Policy During the 1990s." *Current History*, March:105–32.

Kahn, Joseph. 2001. "US Opposes Plan for Financing of Clean Energy over Fossil Fuel." *New York Times*, July 14.

Kamieniecki, Sheldon. 1991. "Political Mobilization, Agenda Building and International Environmental Policy." *Journal of International Affairs* 44(2):339–59.

Kapstein, Ethan. 1990. *The Insecure Alliance: Energy Crises and Western Politics since 1944*. Oxford: Oxford University Press.

Karshenas, Massoud. 1990. *Oil, State and Industrialization in Iran*. Cambridge: Cambridge University Press.

Katouzian, Homa. 2000. *State and Society in Iran*. London: I. B. Tauris.

Kaur, Amarjit. 1990. "Hewers and Haulers." *Modern Asian Studies* 24(1):75–113.

Kazuo, Nimura. 1997. *The Ashio Riot of 1907: A Social History of Mining in Japan*. Durham, NC: Duke University Press.

Kealey, Gregory. 1995. *Workers and Canadian History*. Montreal: McGill-Queen's University Press.

Keck, Margaret, and Kathryn Sikkink. 1998. *Activists beyond Borders: Advocacy Networks in International Politics*. Ithaca, NY: Cornell University Press.

Keddie, Nikki. 2003. *Modern Iran: Roots and Results of Revolution*. New Haven, CT: Yale University Press.

Kemp, Rene. 1994. "Technology and the Transition to Environmental Sustainability." *Futures* 26(10):1023–46.

Kennedy, Paul. 1987. *The Rise and Fall of the Great Powers*. New York: Random House.

Keohane, Robert. 1984. *After Hegemony: Cooperation and Discord in the World Political Economy*. Princeton, NJ: Princeton University Press.

Kerr, Clark, and Abraham Siegel. 1964. "The Interindustry Propensity to Strike—An International Comparison." In *Labor and Management in Industrial Society*, edited by Clark Kerr, 15–36. Garden City, NY: Anchor Books.

Khagram, Sanjeev. 2003. "Neither Temples nor Tombs: A Global Analysis of Large Dams." *Environment* 45(4):28–37.

Kimura, Mitsuhiko. 1995. "The Economics of Japanese Imperialism in Korea, 1910–1939." *Economic History Review* 48(3):555–74.

Kinzer, Stephen. 1999. "The Caspian Accord." *New York Times*, Nov. 19.

Kipp, Jacob. 1976. "Imperial Russia." In *Naval Technology and Social Modernization in the Nineteenth Century*, edited by Ken Hagan, B. Cooling, Jacob Kipp, Bruce Swanson, and A. McMahon. Manhattan, KN: Military Affairs.

Klare, Michael. 2004. *Blood and Oil: The Dangers and Consequences of America's Growing Petroleum Dependency*. New York: Metropolitan Books.

Klinkenborg, Verlyn. 2003. "Turning Northeast Wyoming Upside Down in the Hunt for Coal-Bed Methane." *New York Times*, Dec. 1.

Knape, John. 1987. "British Foreign Policy in the Caribbean Basin 1938–1945." *Journal of Latin American Studies* 19(2):279–94.

Kobrin, Stephen. 1993. "Diffusion as an Explanation of Oil Nationalization." In *Transnational Corporations and the Exploitation of Natural Resources*, edited by Bruce McKern, 284–311. London: Routledge.

Kocka, Jurgen. 1981. "Capitalism and Bureaucracy in German Industrialization before 1914." *Economic History Review (New Series)* 34(3):453–68.

Komanoff, Charles, and Cora Roelofs. 1992. *Fiscal Fission: The Economic Failure of Nuclear Power*. New York: Komanoff Energy Associates and Greenpeace.

Kopp, Otto. 1995. "Fossil Fuels: Coal." *Encyclopedia Brittanica*, pp. 604–12.

Koppel, Tom. 1999. *Powering the Future: The Ballard Fuel Cell and the Race to Change the World*. Toronto: John Wiley & Sons Canada, Ltd.

Kovarik, William. 2001. *Environmental History Timeline*. Radford, VA: Radford University.

Krasner, Stephen. 1978. *Defending the National Interest: Raw Materials Investments and U.S. Foreign Policy*. Princeton, NJ: Princeton University Press.

Kryza, Frank. 2003. *The Power of Light: The Epic Story of Man's Quest to Harness the Sun*. New York: McGraw-Hill.

Kumar, Priti. 1997. "Death in the Air." *Down to Earth*, Nov. 15.

Lambert, Andrew. 1996. "The British Naval Strategic Revolution, 1815–1854." In *Shipping, Technology and Imperialism*, edited by Gordon Jackson and David Williams. Hants, UK: Scolar Press.

Landes, David. 1969. *The Unbound Prometheus*. Cambridge: Cambridge University Press.

Laux, James. 1963. "Some Notes on Entrepreneurship in the Early French Automobile Industry." *French Historical Studies* 3(1):129–34.

Lavigne, Marie. 1983. "The Soviet Union inside Comecon." *Soviet Studies* 35(2):135–53.

Law, Alex. 1999. "Fuel Cells Become Clean Air Apparent for Auto Industry." *Toronto Star*, April 24.

Lee, W. 1988. "Economic Development and the State in Nineteenth Century Germany." *Economic History Review (New Series)* 41(3):346–67.

Leitmann, Josef. 1999. *Sustaining Cities*. New York: McGraw Hill.

Lenczowski, George. 1960. *Oil and State in the Middle East*. Ithaca, NY: Cornell University Press.

Leslie, Stuart. 1993. *The Cold War and American Science*. New York: Columbia University Press.

Levine, Solomon. 1958. *Industrial Relations in Postwar Japan*. Urbana, IL: University of Illinois Press.

Lewis, Gwynne. 1993. *The Advent of Modern Capitalism in France, 1770–1840*. Oxford: Clarendon Press.

Lieber, Robert. 1979. "Europe and America in the World Energy Crisis." *International Affairs* 55(4):531–45.

Lieuwen, Edwin. 1985. "The Politics of Energy in Venezuela." In *Latin American Oil Companies and the Politics of Energy*, edited by John Wirth, 189–225. Lincoln, NB: University of Nebraska Press.

Lipschutz, Ronnie. 1989. *When Nations Clash: Raw Materials, Ideology, and Foreign Policy*. New York: Ballinger Publishing Company.

Llanos, Miguel. 2002. "US Maps Path to Hydrogen Economy." *MSNBC*, Nov. 13.

Long, Bill. 1997. "Environmental Regulation." *OECD Observer* 206:14–18.

Long, Priscilla. 1989. *Where the Sun Never Shines: A History of America's Bloody Coal Industry*. New York: Paragon House.

Looney, Robert, and David Winterford. 1995. *Economic Causes and Consequences of Defense Expenditures in the Middle East and South Asia.* Boulder, CO: Westview Press.

Lovins, Amory. 1977. *Soft Energy Paths: Toward a Durable Peace.* San Francisco: Friends of the Earth.

Lowen, Rebecca. 1987. "Entering the Atomic Power Race: Science, Industry, and Government." *Political Science Quarterly* 102(3):459–79.

Loyola-Diaz, Rafael. 1991. *El ocaso del radicalismo revolucionario: ferrocarrileros y petroleros.* Mexico City: Instituto de Investigaciones Sociales.

Luke, Timothy. 1983. "Dependent Development and the Arab OPEC States." *Journal of Politics* 45(4):979–1003.

Lustick, I. 1997. "The Absence of Middle Eastern Great Powers." *International Organization* 51(4):653–83.

Lutzenhiser, Loren. 1993. "Social and Behavioral Aspects of Energy Use." *Annual Review of Energy and the Environment* 18:247–89.

Lyon, Thomas. 1990. "Natural Gas Policy: The Unresolved Issues." *Energy Journal* 11(2):23–49.

Machlup, Fritz, and Edith Penrose. 1950. "The Patent Controversy in the Nineteenth Century." *Journal of Economic History* 10(1):1–29.

Maddison, Angus. 1995. *Monitoring the World Economy 1820–1992.* Paris: OECD Development Centre.

Magraw, Roger. 1992. *A History of the French Working Class.* Oxford: Blackwell.

Mandel, Ernest. 1980. *Long Waves of Capitalist Development: The Marxist Interpretation.* Cambridge: Cambridge University Press.

Mann, Michael. 1988. *States, War, and Capitalism.* New York: Basil Blackwell.

Marland, Gregg, Tom Boden, Antoinette Brenkert, Bob Andres, and Cathy Johnston. 1999. *Global CO2 Emissions from Fossil-Fuel Burning, Cement Manufacture, and Gas Flaring: 1751–1996.* Oak Ridge, TN: Carbon Dioxide Information Analysis Center. Oak Ridge National Laboratory.

Marples, David. 1997. "Nuclear Power in the Former USSR." In *Nuclear Energy and Security in the Former Soviet Union,* edited by David Marples and Marilyn Young. Boulder, CO: Westview Press.

Marquis, Christopher. 2003. "Rumsfeld Made Iraq Overture in '84 Despite Chemical Raids." *New York Times,* Dec. 23.

Marsh, Charles. 1928. *Trade Unionism in the Electric Light and Power Industry.* Urbana, IL: University of Illinois Press.

Martin, Jean-Marie. 1996. "Energy Technologies: Systemic Aspects, Technological Trajectories, and Institutional Frameworks." *Technological Forecasting and Social Change* 53:81–95.

Marwah, Onkar. 1977. "India's Nuclear and Space Programs." *International Security* 2(2):96–121.

Marx, Karl. 1967. *Capital, Vol 1.* New York: International Publishers.

Marx, Karl, and Friedrich Engels. 1848. *The Manifesto of the Communist Party.* Moscow: Progress Publishers.

Masters, Charles, Emil Attanasi, and David Root. 1994. *World Petroleum Assessment and Analysis*. Reston, VA: U.S. Geological Survey.

Mathias, Peter. 1969. *The First Industrial Nation*. New York: Methuen.

Mathieson, R. 1980. "Nuclear Power in the Soviet Bloc." *Annals of the Association of American Geographers* 70(2):271–79.

Maycock, Paul. 1984. "U.S.-Japanese Competition for the World Photovoltaic Market." In *U.S.-Japanese Energy Relations: Cooperation and Competition*, edited by Charles Ebinger and Ronald Morse. Boulder, CO: Westview Press.

McCahill, Michael. 1976. "Peers, Patronage, and the Industrial Revolution, 1760–1800." *Journal of British Studies* 16(1):84–107.

McCully, Patrick. 1996. *Silenced Rivers: The Ecology and Politics of Large Dams*. London: Zed Books.

McFarland, Marvin. 1990. *The Papers of Wilbur and Orville Wright*. Salem, N.H.: Ayer Company Publishers.

McGeary, Johanna. 2004. "Terror's Next Wave." *Time*, March 29:24–6.

McGowan, Jon, and Stephen Connors. 2000. "Windpower." *Annual Review of Energy and the Environment* 25:147–97.

McKern, Bruce. 1996. "Transnational Corporations and the Exploitation of Natural Resources." In *Transnational Corporations and World Development*. London: International Thompson Business Press.

McLauchlan, Gregory. 1997. "World War II and the Transformation of the US State." *Sociological Inquiry* 67(1):1–26.

McLaughlin, Charles. 1954. "The Stanley Steamer: A Study in Unsuccessful Innovation." *Explorations in Entrepreneurial History* 7:37–47.

McNeill, William. 1982. *The Pursuit of Power*. Chicago, IL: University of Chicago Press.

Melamid, Alexander. 1994. "International Trade in Natural Gas." *Geographical Review* 84(2):216–22.

Melby, Eric. 1981. *Oil and the International System: The Case of France, 1918–1969*. Baltimore, MD: Arno Press.

Meyer, John, David Frank, Ann Hironaka, Evan Schofer, and Nancy Tuma. 1997. "The Structuring of a World Environmental Regime, 1870–1990." *International Organization* 51(4):623.

Meyers, Stephen, and Lee Schipper. 1992. "World Energy Use in the 1970s and 1980s." *Annual Review of Energy and the Environment* 17:463–505.

Michel, Joel. 1990. "Industrial Relations in French Coal Mining." In *Workers, Owners and Politics in Coal Mining*, edited by Gerald Feldman and Klaus Tenfelde, 271–314. New York: St. Martin's Press.

Miguel, Antonio. 1991. "Environmental Pollution Research in South America." *Environmental Science and Technology* 25(4):590–93.

Milner, Simon, and David Metcalf. 1991. *A Century of UK Strike Activity*. London: Center for Economic Performance, London School of Economics.

Milward, Alan. 1977. *War, Economy and Society: 1939–1945*. Berkeley, CA: University of California Press.

Minami, Ryoshin. 1977. "Mechanical Power in the Industrialization of Japan." *Journal of Economic History* 37(4):935–58.

Minor, Michael. 1994. "The Demise of Expropriation as an Instrument of LDC Policy, 1980–1992." *Journal of International Business Studies* 25(1): 177–188.

Mitchell, Brian. 1982. *International Historical Statistics: Africa and Asia*. New York: New York University Press.

———. 1983. *International Historical Statistics: The Americas and Australasia*. Detroit, MI: Gale Research Company.

———. 1984. *Economic Development of the British Coal Industry, 1800–1914*. Cambridge: Cambridge University Press.

———. 1988. *British Historical Statistics*. Cambridge: Cambridge University Press.

———. 1992. *International Historical Statistics: Europe, 1750–1988*. New York: Stockton Press.

———. 1993. *International Historical Statistics: The Americas 1750–1988*. New York: Stockton Press.

Moaddel, Mansoor. 1991. "Class Struggle in Post-Revolutionary Iran." *International Journal of Middle Eastern Studies* 23:317–43.

Mock, John, Jefferson Tester, and P. Wright. 1997. "Geothermal Energy from the Earth." *Annual Review of Energy and the Environment* 22:305–57.

Modelski, George, and William Thompson. 1996. *Leading Sectors and World Powers*. Columbia, SC: University of South Carolina Press.

Moghadam, Valentine. 1988. "Industrialization Strategy and Labour's Response: The Case of the Workers' Councils in Iran." In *Trade Unions and the New Industrialization of the Third World*, edited by Roger Southall, 182–209. Pittsburgh, PA: University of Pittsburgh Press.

Montgomery, David. 1980. "Strikes in Nineteenth-Century America." *Social Science History* 4(1):81–104.

Moore, Jason. 2003. "The Modern World-System as Environmental History? Nature and the Future of World-Systems Analysis." *Theory and Society* 32:307–77.

Moore, Taylor. 1997. "Market Potential High for Fuel Cells." *EPRI Journal* 22(3):6–17.

Moran, Theodore. 1987. "Managing an Oligopoly of Would-Be Sovereigns." *International Organization* 41(4):575–607.

Morse, Ronald. 1992. "Japan: Crafting an Energy Strategy for Competitiveness in the World Market." *Harvard International Review* 14(2):14–17.

Moskoff, William. 1973. "The USSR and Developing Countries: Politics and Export Prices, 1955–69." *Soviet Studies* 24(3):348–63.

Murphy, Raymond, and Hugh Spittal. 1945. "Movements of the Center of Coal Mining in the Appalachian Plateaus." *Geographical Review* 35(4):624–33.

Myerscough, John. 1985. "Airport Provision in the Inter War Years." *Journal of Contemporary History* 20(1):41–70.

Nakicenovic, Nebojsa, Arnulf Grubler, and Alan McDonald. 1998. *Global Energy Perspectives*. Cambridge: Cambridge University Press.

Nash, Gerald. 1968. *United States Oil Policy, 1890–1964*. Pittsburgh, PA: University of Pittsburgh Press.

National Academy of Sciences. 2002. *Abrupt Climate Change: Inevitable Surprises*. Washington DC: National Academy Press.

———. 2004. *The Hydrogen Economy: Opportunities, Costs, Barriers, and R&D Needs*. Washington DC: National Academies Press.

National Energy Policy Development Group. 2001. *National Energy Policy*. Washington, DC: White House.

National Renewable Energy Laboratory. 2002. *Renewable Energy Cost Trends*. Golden, CO: U.S. NREL.

National Security Archive. 2003. *Shaking Hands with Saddam Hussein: The U.S. Tilts toward Iraq, 1980–1984*. Washington, DC: National Security Archive.

Naus, Donald. 1999. "Taking Fuel-Cell Technology Further." *Los Angeles Times*, March 18.

Navarro, Peter. 1983. "Union Bargaining Power in the Coal Industry, 1945-1981." *Industrial and Labor Relations Review* 36(2):214–29.

Navickis, Roberta. 1978. "Biomass." *Science News* 113(16):258–60.

Needham, Joseph. 1964. *The Development of Iron and Steel Technology in China*. Cambridge, MA: W. Heffer and Sons.

Neufeld, John. 1987. "Price Discrimination and the Adoption of Electricity Demand Charge." *Journal of Economic History* 47(3):693–709.

Nilsson, Lars, and Thomas Johansson. 1994. "Environmental Challenges to the Energy Industries." In *Sustainable Development and the Energy Industries: Implementation and Impacts of Environmental Legislation*, edited by Nicola Steen, 47–79. London: Earthscan Publications.

Nitzan, Jonathan, and Shimshon Bichler. 1995. "Bringing Capital Accumulation Back In." *Review of International Political Economy* 2(3):446–515.

Noorani, A. 1981. "Indo-US Nuclear Relations." *Asian Survey* 21(4):399–416.

Nore, Petter. 1980. "Oil and the State." In *Oil and Class Struggle*, edited by Petter Nore and Terisa Turner, 69–88. London: Zed Books.

Nowell, Gregory. 1994. *Mercantile States and the World Oil Cartel, 1900–1939*. Ithaca, NY: Cornell University Press.

Nuclear Energy Institute. 2003. *Nuclear Data*. Washington, DC: NEI.

Nye, David. 1990. *Electrifying America: Social Meanings of a New Technology, 1880–1940*. Cambridge, MA: MIT Press.

———. 1998. *Consuming Power: A Social History of American Energies*. Cambridge, MA: MIT Press.

———. 1999. "Path Insistence: Comparing European and American Attitudes toward Energy." *Journal of International Affairs* 53(1):129–48.

Nye, Joseph. 1981. "Japan." In *Energy and Security*, edited by David Deese and Joseph Nye. Cambridge, MA: Ballinger Publishing Company.

O'Riordan, Timothy, and William Clark. 1995. "The Legacy of Earth Day." *Environment* 37(3):6–22.

Organization for Economic Cooperation and Development. 1993. *Environmental Policies and Industrial Competitiveness*. Paris: OECD.

Ogden, Joan. 1999. "Prospects for Building a Hydrogen Energy Infrastructure." *Annual Review of Energy and the Environment* 24:227–79.

Ogden, Joan, Robert Williams, and Eric Larson. 2001. *Toward a Hydrogen-Based Transportation System*. Princeton, NJ: Center for Energy and Environmental Studies, Princeton University.

Ogutcu, Mehmet. 1998. "China and the World Energy System." *Journal of Energy and Development* 23(2):281–318.

Okruhlik, Gwenn. 1999. "Rentier Wealth, Unruly Law, and the Rise of Opposition." *Comparative Politics* 31(3):295–315.

Olien, Roger. 2002. "Oil and Natural Gas Industry." In *The New Handbook of Texas*, edited by Douglas Barnett. Austin, TX: Texas State Historical Association.

Ostwald, W. 1909. *Energetische Grundlagen der Kulturwissenschaften*. Leipzig: Vorvort.

Overy, R. 1973. "Transportation and Rearmament in the Third Reich." *Historical Journal* 16(2):389–409.

Owen, Edgar. 1975. *Trek of the Oil Finders*. Tulsa, OK: American Association of Petroleum Geologists.

Painter, David. 1986. *Oil and the American Century: The Political Economy of U.S. Foreign Oil Policy, 1941–1954*. Baltimore, MD: Johns Hopkins University Press.

Palcan Fuel Cells. 2003. "Palcan Fuel Cells Announces a Joint Venture." *Future Energies*, Dec. 3.

Palladino, Grace. 1990. *Another Civil War: Labor, Capital, and the State in the Anthracite Regions of Pennsylvania 1840-68*. Urbana, IL: University of Illinois Press.

Paris, Michael. 1991. "The First Air Wars." *Journal of Contemporary History* 26(1):97–109.

Parnell, Martin. 1994. *The German Tradition of Organized Capitalism*. Oxford: Clarendon Press.

Parsa, Misagh. 1988. "Theories of Collective Action and the Iranian Revolution." *Sociological Forum* 3(1):44–71.

Pastor, Manuel. 1989. "Latin America, the Debt Crisis, and the International Monetary Fund." *Latin American Perspectives* 16(1):79–110.

Patterson, Walter. 1990. "Thatcher's Failed Romance with Nuclear Power." *Bulletin of the Atomic Scientists* 46(3):1–3.

Paxson, Frederic. 1946. "The Highway Movement, 1916–1935." *American Historical Review* 51(2):236–53.

Payne, Peter. 1978. "Industrial Entrepreneurship and Management in Great Britain." In *The Cambridge Economic History of Europe*, edited by Peter Mathias and M. Postan, 180–230. Cambridge: Cambridge University Press.

Pearce, Fred. 2002. "Tree Farms Won't Halt Climate Change." *New Scientist Online News*, Oct. 28.

Penrose, Edith. 1968. *The Large International Firms in Developing Countries: The International Petroleum Industry*. London: Allen & Unwin.

———. 1987. "The Structure of the International Oil Industry." In *The International Oil Industry: An Interdisciplinary Perspective*, edited by Judith Rees and Peter Odell. New York: St. Martin's Press.

———. 1993. "The Nature and Economic Significance of the Large International Firms." In *Transnational Corporations and the Exploitation of Natural Resources*, edited by Bruce McKern, 37–68. London: Routledge.

Perelman, Lewis. 1981. "Speculations on the Transition to Sustainable Energy." In *Energy Transitions: Long-Term Perspectives*, edited by Lewis Perelman, August Giebelhaus, and Michael Yokell. Boulder, CO: Westview Press.

Perez-Pena, Richard. 1996. "Quiet Boomlet in Vehicles Using Compressed Natural Gas a Fuel." *New York Times*, July 15:B1.

Perlin, John. 1989. *A Forest Journey: The Role of Wood in the Development of Civilization*. New York: W.W. Norton & Company.

Perlin, John, and Ken Butti. 2002. *Solar Evolution: The History of Solar Energy*. Santa Barbara, CA: The Rahus Institute.

Perrone, Luca. 1983. "Positional Power and Propensity to Strike." *Politics & Society* 12(2):231–61.

———. 1984. "Positional Power, Strikes and Wages." *American Sociological Review* 49(3):412–26.

Peterson, Florence. 1938. *Review of Strikes in the United States*. Washington, DC: U.S. Bureau of Labor Statistics.

Peterson, John. 2002. *Saudi Arabia and the Illusion of Security*. London: Oxford University Press.

Pew Research Center. 2004. *A Year After the Iraq War*. Washington, DC: Pew Research Center for the People and the Press.

Philip, George. 1982. *Oil and Politics in Latin America: Nationalist Movements and State Companies*. Cambridge: Cambridge University Press.

Phimister, Ian. 1994. *Wangi Kolia: Coal, Capital and Labour in Colonial Zimbabwe 1894–1954*. Johanesburg: Witwatersrand University Press.

Plotkin, Steven. 1980. "Energy from Biomass." *Environment* 22(9):6–18.

Polanyi, Karl. 1957. *The Great Transformation*. Boston: Beacon Press.

Pollard, Sidney. 1965. *The Genesis of Modern Management*. Cambridge, MA: Harvard University Press.

———. 1978. "Labour in Great Britain." In *The Cambridge Economic History of Europe*, edited by Peter Mathias and M. Postan, 97–180. Cambridge: Cambridge University Press.

———. 1980. "A New Estimate of British Coal Production, 1750–1850." *Economic History Review* 33(2):212–35.

———. 1994. "Entrepreneurship, 1870–1914." In *The Economic History of Britain Since 1700, Volume 2*, edited by Roderick Floud and Donald McCloskey, 62–89. Cambridge: Cambridge University Press.

Pollard, Sidney, and Paul Robertson. 1979. *The British Shipbuilding Industry, 1870–1914*. Cambridge, MA: Harvard University Press.

Porter, Gareth, and Janet Brown. 1996. *Global Environmental Politics*. Boulder, CO: Westview Press.

Pratt, Joseph. 1981. "The Ascent of Oil." In *Energy Transitions: Long-Term Perspectives*, edited by Lewis Perelman, August Giebelhaus and Michael Yokell. Boulder, CO: Westview Press.

Price, Roger. 1988. *The Revolutions of 1848*. London: MacMillan.

Priest, Dana. 2003. "Rumsfeld Visited Baghdad in 1984 to Reassure Iraqis." *Washington Post*, Dec. 19.

Pycroft, Christopher, and Barry Munslow. 1988. "Black Mine Workers in South Africa: Strategies of Co-option and Resistance." *Journal of Asian and African Studies* 23(2):156–79.

Quinn, Andrew. 2002. "Data Shows World Awash in Stolen Nuclear Material." *Reuters News Service*, March 7.

Rae, John. 1967. "Energy Conversion." In *Technology in Western Civilization, Volume I*, edited by Melvin Kranzberg and Carroll Pursell. New York: Oxford University Press.

Ramazani, R. 1980. "Iran's Revolution." *International Affairs* 56(3):443–57.

Redlich, Fritz. 1944. "The Leaders of the German Steam-Engine Industry during the First Hundred Years." *Journal of Economic History* 4(2):121–48.

Reich, Michael. 1983. "Environmental Policy and Japanese Society." *International Journal of Environmental Studies* 20:199–207.

Reid, Donald. 1985. *The Miners of Decazeville*. Cambridge, MA: Harvard University Press.

Reisinger, William. 1992. *Energy and the Soviet Bloc*. Ithaca, NY: Cornell University Press.

Renner, Michael. 1984. "Restructuring the World Energy Industry." *MERIP Reports* (120):12–31.

Reuters. 2001. "Asia Makes Big Push into Clean, Alternative Fuels." *Reuters News Service*, June 20.

Rhodes, Richard. 1988. *The Making of the Atomic Bomb*. New York: Simon & Shuster.

Rich, Bruce. 1994. *Mortgaging the Earth*. Boston: Beacon Press.

Rifkin, Jeremy. 2002. *The Hydrogen Economy: The Creation of the World-Wide Energy Web and the Redistribution of Power on Earth*. New York: Jeremy P. Tarcher/Putnam.

Rimlinger, G. 1989. "Labour and the State on the Continent, 1800–1939." In *The Cambridge Economic History of Europe, Volume VIII*, edited by Peter Mathias and Sidney Pollard, 549–606. Cambridge: Cambridge University Press.

Riposa, Gerry. 1988. "After Reagan's Deregulation." *Policy Studies Review* 8(1): 36–54.

Riva, Joseph, and Gordon Atwater. 2002. "Natural Gas." *Encyclopedia Britannica*.

Robinson, Eric. 1974. "The Early Diffusion of Steam Power." *Journal of Economic History* 34(1):91–106.

Robinson, R. 1972. "Non-European Foundations of European Imperialism." In *Studies in the Theory of Imperialism*, edited by R. Owen and B. Sutcliffe, 115–25. London: Longman.

Rohrschneider, R. 1991. "Public Opinion toward Environmental Groups in Western Europe." *Social Science Quarterly* 72(2):251–67.

Romer, Alfred. 1982. *The Restless Atom*. New York: Dover.

Romm, Joseph. 2004. *The Hype about Hydrogen*. Washington, DC: Island Press.

Roncaglia, Alessandro. 1983. *The International Oil Market*. London: MacMillan.

Rosa, Eugene, and Riley Dunlap. 1994. "Nuclear Power: Three Decades of Public Opinion." *Public Opinion Quarterly* 58:295–325.

Rosenberg, Nathan. 1994. *Exploring the Black Box*. New York: Cambridge University Press.

Rothstein, Linda. 2003. "Goodness Gracious, Great Balls of Power." *Bulletin of the Atomic Scientists* 59(5):10–11.

Royal Commission. 1926. *Report of the Royal Commission on the Coal Industry*. London: H.M. Stationery Office.

Rubinstein, Alvin. 1981. "The Soviet Union and Iran under Khomeini." *International Affairs* 57(4) :599–617.

Rucht, Dieter. 1995. "The Impact of Anti-Nuclear Power Movements in International Comparison." In *Resistance to New Technology*, edited by Martin Bauer, 277–91. Cambridge: Cambridge University Press.

Rudig, Wolfgang. 1990. *Anti-Nuclear Movements: A World Survey of Opposition to Nuclear Energy*. London: Longman Group.

Rudig, Wolfgang. 1991. "Green Party Politics around the World." *Environment* 33(8):6–17.

Russett, Bruce. 1985. "The Mysterious Case of Vanishing Hegemony." *International Organization* 39(2):207–31.

Sabel, Charles, and Jonathan Zeitlin. 1985. "Historical Alternatives to Mass Production." *Past and Present* (108):133–76.

Said, Edward. 1994. *The Politics of Dispossession*. New York: Pantheon Books.

Sampson, Anthony. 1975. *The Seven Sisters: The Great Oil Companies and the World They Made*. New York: Viking Press.

Samuels, Richard. 1987. *The Business of the Japanese State*. Ithaca, NY: Cornell University Press.

Sandbach, F. 1978. "A Further Look at the Environment as a Political Issue." *International Journal of Environmental Studies* 12(2):99–111.

Schipper, Lee, Fridtjof Unander, Scott Murtishaw, and Mike Ting. 2001. "Indicators of Energy Use and Carbon Emissions." *Annual Review of Energy and the Environment* 26:49–81.

Schmidheiny, Stephan, and Federico Zorraquin. 1996. *Financing Change*. Cambridge, MA: MIT Press.

Schmitz, Christopher. 1995. "The World's Largest Industrial Companies of 1912." *Business History* 37(4):85–94.

Schoenberger, Erica. 1997. *The Cultural Crisis of the Firm*. Cambridge, MA: Blackwell.

Schowengerdt, Stephen. 1999. "Solar Cell Efficiency Makes Big Leaps." *Environmental News Network*, Oct. 27.

Schumpeter, Joseph. 1949. *The Theory of Economic Development*. Cambridge, MA: Harvard University Press.

Schurr, Sam, Calvin Burwell, Warren Devine, and Sidney Sonenblum. 1990. *Electricity in the American Economy*. New York: Greenwood Press.

Schurr, Samuel, and Bruce Netschert. 1960. *Energy in the American Economy, 1850–1975*. Baltimore, MD: Johns Hopkins University Press.

Schwartz, Peter, and Doug Randall. 2003. *An Abrupt Climate Change Scenario and Its Implications for United States National Security*. Emeryville, CA: Global Business Network (Report Commissioned by the U.S. Department of Defense).

Seltzer, Curtis. 1985. *Fire in the Hole*. Lexington, MA: University of Kentucky Press.

Sen, Sunil. 1994. *Working Class Movements in India, 1885–1975*. New Delhi: Oxford University Press.

Sharlin, Harold. 1967. "Applications of Electricity." In *Technology in Western Civilization, Volume I*, edited by Melvin Kranzberg and Carroll Pursell. New York: Oxford University Press.

Shepherd, Robert. 1993. *Ancient Mining*. London: Elsevier Applied Science.

Shorter, Edward, and Charles Tilly. 1974. *Strikes in France 1830–1968*. Cambridge: Cambridge University Press.

Showalter, Dennis. 1975. *Railroads and Rifles: Soldiers, Technology, and the Unification of Germany*. Hamden, CN: Archon Books.

Sieferle, Rolf. 2001. *The Subterranean Forest: Energy Systems and the Industrial Revolution*. Cambridge, MA: White Horse Press.

Siegelbaum, Lewis, and Daniel Walkowitz. 1995. *Workers of the Donbass Speak*. Albany, NY: State University of New York Press.

Siemens AG. 1999. "Shell and Siemens to Develop Emission-Free Fuel Cell Power Plant." *Press Release*, July 13.

Siemens. 2002. *History of Siemens and Halske*. Berlin: Simens Corporation.

Silver, Beverly. 2003. *Forces of Labor: Workers' Movements and Globalization Since 1870*. Cambridge: Cambridge University Press.

Silver, Beverly, and Eric Slater. 1999. "The Social Origins of World Hegemonies." In *Chaos and Governance in the Modern World System*, edited by Giovanni Arrighi, Beverly Silver, Terence Hopkins, and Iftikhar Ahmad, 151–240. St. Paul, MN: University of Minnesota Press.

Silverfarb, Daniel. 1996. "The Revision of Iraq's Oil Concession." *Middle Eastern Studies* 32(1):69–95.

Simonson, G. 1960. "The Demand for Aircraft and the Aircraft Industry, 1907–1958." *Journal of Economic History* 20(3):361–82.

Sirageldin, I. 1988. "Future Arab Economic Potential." In *The Next Arab Decade: Alternative Futures*, edited by Hisham Sharabi, 185–207. Boulder, CO: Westview Press.

Skocpol, Theda. 1979. *States and Social Revolutions: A Comparative Analysis of France, Russia and China*. Cambridge: Cambridge University Press.

Smil, Vaclav. 1999. *Energies*. Cambridge, MA: MIT Press.

———. 2003. *Energy at the Crossroads: Global Perspectives and Uncertainties*. Cambridge, MA: MIT Press.

Smith, Gar. 1990. "The Rise of the US Oil-Ogarchy." *Earth Island Journal* 5(4): 28–30.

Spring, David. 1951. "The English Landed Estate in the Age of Coal and Iron." *Journal of Economic History* 11(1):3–24.

Stearns, Peter. 1993. *The Industrial Revolution in World History*. Boulder, CO: Westview Press.

Stein, Arthur. 1984. "The Hegemon's Dilemma: Great Britain, the United States, and the International Economic Order." *International Organization* 38(2):355–86.

Stivers, William. 1982. *Supremacy and Oil: Iraq, Turkey, and the Anglo-American World Order, 1918–1930*. Ithaca, NY: Cornell University Press.

Stone, Richard. 2002. "Counting the Cost of London's Killer Smog," *Science*, 298(5601): 2106–8.

Sullivan, Eileen. 1983. "Liberalism and Imperialism: J. S. Mill's Defense of the British Empire." *Journal of the History of Ideas* 44(4):599–617.

Supple, Barry. 1987. *The History of the British Coal Industry, Volume 4*. New York: Clarendon Press.

Tainter, Joseph. 1990. *The Collapse of Complex Societies*. Cambridge: Cambridge University Press.

Takagi, Jinzaburo. 1996. "Japan's Plutonium Program." In *Japan's Nuclear Future: The Plutonium Debate and East Asian Security*, edited by Selig Harrison. Washington, DC: Carnegie Endowment for International Peace.

Takao, Yasuo. 2001. "The Rise of the 'Third Sector' in Japan." *Asian Survey* 41(2):290–310.

Talalay, Michael. 1997. "Fuel Cells, Cars, and World Power." In *Technology, Culture and Competitiveness: Change in the World Political Economy*, edited by Michael Talalay, Chris Farrands, and Roger Tooze. New York: Routledge.

Tavoulareas, E., and Jean Charpentier. 1995. *Clean Coal Technologies for Developing Countries*. Washington, DC: World Bank.

Taylor, A. 1968. "The Coal Industry." In *The Development of British Industry and Foreign Competition, 1875–1914*, edited by Derek Aldcroft, 37–70. London: Allen and Unwin.

Teece, David. 1990. "Structure and Organization of the Natural Gas Industry." *Energy Journal* 11(3):1–36.

Temple, Robert. 1986. *The Genius of China*. New York: Simon & Schuster.

Temples, James. 1980. "The Politics of Nuclear Power." *Political Science Quarterly* 95(2):239–60.

Thomas, Brinley. 1986. "Was There an Energy Crisis in Great Britain in the 17th Century?" *Explorations in Economic History* 23:124–52.

Tilly, Charles. 1984. *Big Structures, Large Processes, Huge Comparisons*. New York: Russell Sage Foundation.

Tokumoto, Tom. 1995. "Fuel Cell Development and Commercialization in Japan." In *The Strategic Value of Fossil Fuels: Challenges and Responses*, 235–41. Paris: International Energy Agency.

Tonn, Bruce, and Sujit Das. 2002. "Assessment of Platinum Availability for Advanced Fuel-Cell Vehicles." *Transportation Research Record* (1815):99–104.

Trebilcock, Clive. 1993. "Science, Technology and the Armaments Industry in the UK and Europe." *Journal of European Economic History* 22(3):565–78.

Tromp, Tracey, Run-Lie Shia, Mark Allen, John Eiler, and Y Yung. 2003. "Potential Environmental Impact of a Hydrogen Economy on the Stratosphere." *Science* 300(June 13):1740–42.

Troxel, C. 1937. "Corporate Control in the Natural Gas Industry." *Journal of Business of the University of Chicago* 10(2):147–73.

Tsutsumi, K. 1998. "A Case Study on the Former Coal-Mining Region of Takashima." In *Sustainability and Development: On the Future of Small Society in a Dynamic Economy*, edited by Lennart Andersson and Thomas Blom, 230–5. Karlstad, Sweden: University of Karlstad Publications.

Turner, Louis. 1983. *Oil Companies in the International System*. London: Allen & Unwin.

Uchida, Hoshimi. 1995. *A Short History of Japanese Technology*. Tokyo: The History of Technology Library.

United Nations. 1952. *World Energy Supplies in Selected Years, 1929–1950*. New York: United Nations.

U.S. Bureau of Labor Statistics. 1937. *Strikes in the United States, 1880–1936*. Washington, DC: U.S. Bureau of Labor Statistics.

———. 1950. *Employment Outlook in Petroleum Production and Refining*. Washington, DC: U.S. Bureau of Labor Statistics.

———. 1979. *Technological Change and its Labor Impact in Five Energy Industries*. Washington, DC: U.S. Bureau of Labor Statistics.

———. 1998 (and other years). *Analysis of Work Stoppages*. Washington, DC: U.S. Bureau of Labor Statistics.

U.S. Department of Commerce. 1975. *Historical Statistics of the United States*. Washington, DC: Department of Commerce.

U.S. Department of Energy. 2002. *Turning Sunlight into Electricity*. Washington, DC: U.S. DOE.

U.S. Department of Labor. 2000. *Injury Trends in Mining*. Washington, DC: Mining Safety and Health Administration.

U.S. Energy Information Administration. 2003. *International Energy Outlook 2003*. Washington, DC: U.S. EIA.

U.S. Environmental Protection Agency. 1997. *Public Health Effects of Ozone and Fine Particle Pollution*. Washington, DC: U.S. EPA.

U.S. Geological Survey. 2000. *World Petroleum Assessment 2000*. Denver, CO: U.S. Geological Survey.

Van Natta, Don, and Desmond Butler. 2003. "Calls to Jihad Are Said to Lure Hundreds of Militants into Iraq." *New York Times*, Oct. 31.

Veenstra, Theodore, and Wilbert Fritz. 1936. "Major Economic Tendencies in the Bituminous Coal Industry." *Quarterly Journal of Economics* 51(1):106–30.

Veiluva, Michael. 1994. "Federal Responsibilities and Realities: An Alternative View of the Cleanup of the Nuclear Weapons Complex." *Social Justice* 22(4):126–39.

Venn, Fiona. 1986. *Oil Diplomacy in the Twentieth Century*. New York: St. Martin's Press.

Vernon, Raymond. 1983. *Two Hungry Giants: The United States and Japan in the Quest for Oil and Ores*. Cambridge, MA: Harvard University Press.

———. 1993. "The Raw Materials Ventures." In *Transnational Corporations and the Exploitation of Natural Resources*, edited by Bruce McKern, 68–93. London: Routledge.

Ville, Simon. 1990. *Transport and the Development of the European Economy, 1750–1918*. London: MacMillan.

Voskuil, Walter. 1942. "Coal and Political Power in Europe." *Economic Geography* 18(3):247–58.

Wald, Matthew. 1999. "Energy to Count On." *New York Times*, Aug. 17.

Wall, Bennett. 1988. *History of Standard Oil Company*. New York: McGraw-Hill.

Wallace, Michael, Larry Griffin, and Beth Rubin. 1989. "The Positional Power of American Labor, 1963–1977." *American Sociological Review* 54:197–214.

Wallerstein, Immanuel. 1974. *The Modern-World System: Capitalist Agriculture and the Origins of the European World-Economy in the Sixteenth Century*. San Diego, CA: Academic Press.

———. 1979. *The Capitalist World-Economy*. Cambridge: Cambridge University Press.

———. 1984. "The Three Instances of Hegemony in the History of the Capitalist World-Economy." In *Current Issues and Research in Macrosociology, International Studies in Sociology and Social Anthropology*, edited Gerhard Lenski, 100–108. Leiden: E. J. Brill.

———. 1992. "America and the World: Today, Yesterday, and Tomorrow." *Theory and Society* 21(1):1–28.

Wapner, Paul. 1996. *Environmental Activism and World Civic Politics*. Albany, NY: State University of New York Press.

Weber, Max. 1947. *The Theory of Social and Economic Organization*. New York: Free Press.

Weidner, Helmut. 1995. *25 Years of Modern Environmental Policy in Germany*. Berlin: Social Science Research Center.

Weisbrod, Bernd. 1990. "Entrepreneurial Politics and Industrial Relations in Mining in the Ruhr Region." In *Workers, Owners and Politics in Coal Mining*, edited by Gerald Feldman and Klaus Tenfelde, 118–202. New York: St. Martin's Press.

Weisman, Alan. 1998. *Gaviotas: A Village to Reinvent the World*. White River Junction, VT: Chelsea Green Publishing Company.

Werner, Herbert. 1963. "Labor Organizations in the American Petroleum Industry." In *The American Petroleum Industry: The Age of Energy, 1899–1959*, edited by Harold Williamson, Ralph Andreano, Arnold Daum, and Gilbert Klose, 827–45. Evanston, IL: Northwestern University Press.

White, Lynn. 1960. "Tibet, India, and Malaya as Sources of Western Medieval Technology." *The American Historical Review* 65(3):515–26.

Wilkins, Mira. 1993. "European and North American Multinationals, 1870–1914." In *Transnational Corporations: A Historical Perspective*, edited by Geoffrey Jones. London: Routledge.

Williams, Chris. 1998. *Capitalism, Community and Conflict: The South Wales Coalfield, 1898–1947*. Cardiff: University of Wales Press.

Williamson, Harold, Ralph Andreano, Arnold Daum, and Gilbert Klose. 1963. *The American Petroleum Industry*. Evanston, IL: Northwestern University Press.

Wilson, Trevor, and Robin Prior. 2001. "Conflict, Technology, and the Impact of Industrialization." *Journal of Strategic Studies* 24(3):128–57.

Winkler, Allan. 1993. "The 'Atom' and American Life." *The History Teacher* 26(3):317–37.

Winterton, Jonathan. 1994. "Social and Technological Characteristics of Coal-Face Work." *Human Relations* 47(1):89–117.

World Bank. 1999. *Fuel for Thought: Environmental Strategy for the Energy Sector*. Washington, DC: World Bank.

World Commission on Dams. 2000. *Dams and Development: A New Framework for Decision-Making*. Cape Town, South Africa: World Commission on Dams.

World Energy Council. 1990. *Report of the Fuel Cells Committee*. Paris: World Energy Council.

———. 1994. *New Renewable Energy Resources*. London: Kogan Page.

———. 2000. *Energy for Tomorrow's World*. London: World Energy Council.

World Meteorological Organization. 2003. *WMO Statement on the Status of the Global Climate in 2002*. Geneva: World Meteorological Organization.

Wright, Tim. 1981. "Growth of the Modern Chinese Coal Industry." *Modern China* 7(3):317–50.

Wynn, Charters. 1992. *Workers, Strikes, and Pogroms*. Princeton, NJ: Princeton University Press.

Yago, Glenn. 1983. "The Sociology of Transportation." *Annual Review of Sociology* 9:171–90.

Yasuba, Yasukichi. 1996. "Did Japan Suffer from a Shortage of Natural Resources Before World War II?" *Journal of Economic History* 56(3):543–60.

Yergin, Daniel. 1991. *The Prize: The Epic Quest for Oil, Money, and Power*. New York: Simon and Schuster.

Yuru, Wang. 1994. "Capital Formation and Operating Profits of the Kailuan Mining Administration, 1903–1937." *Modern Asian Studies* 28(1):99–128.

Index

Bruce Podobnik is Associate Professor of Sociology at Lewis and Clark College. He is co-editor (with Thomas Reifer) of *Transforming Globalization: Challenges and Opportunities in the Post 9/11 Era,* as well as a number of articles on environmental and social movement topics.